Exploring Introductory Algebra with the TI-81®

Charles Vonder Embse
Central Michigan University

Eugene Olmstead
Elmira Free Academy

Addison-Wesley Publishing Company, Inc.
Reading, Massachusetts • Menlo Park, California • New York
Don Mills, Ontario • Wokingham, England • Amsterdam • Bonn
Sydney • Singapore • Tokyo • Madrid • San Juan • Milan • Paris

The programs and applications presented in this book have been included for their instructional value. They have been tested with care but are not guaranteed for any particular purpose. The publisher does not offer any warranties or representations, nor does it accept any liabilities with respect to the programs or applications.

ISBN 0-201-52431-7

1 2 3 4 5 6 7 8 9 10-MU-95949392

Contents

Introduction **1**

1 Numerical Operations and Linear Functions **3**

1.1 Selecting Modes of Operation on the TI-81 3

1.2 Exploring Numerical Calculations and Order of Operations 6

1.3 Exploring the Viewing Rectangle, the Coordinate Plane, Points, and Lines 17

1.4 Exploring Lines and Linear Functions 36

1.5 Exploring Linear Inequalities 49

1.6 Exploring Systems of Linear Equations 65

1.7 Exploring Systems of Linear Inequalities 82

2 Quadratic Equations and Functions **97**

2.1 Exploring Quadratic Equations 97

2.2 Exploring Families of Quadratic Functions 116

2.3 Exploring Problems Represented by Quadratic Functions 133

2.4 Exploring More Families of Functions 146

3 Exploring Polynomial, Rational, and Radical Functions **159**

3.1 Exploring Polynomial Functions 159

3.2 Exploring Rational Functions 182

3.3 Exploring Radical Functions 202

Index **215**

Introduction

The Texas Instruments TI-81 graphics calculator is really a multidimensional, user-friendly tool for exploration, experimentation, and problem solving in the mathematics classroom. Calling this instrument a graphing calculator is a misnomer on two counts. First, it is not simply a calculator. While it will do all of the computational calculations that a "standard" scientific calculator will do, the large presentation screen and the interactive modes of operation make the TI-81 much more than just another scientific calculator. In fact, the TI-81 is probably the best scientific calculator you have ever used. Secondly, this tool is not limited to graphics presentations. The best way to describe the TI-81 is as a computer that fits in the palm of your hand. Like other powerful computers, the TI-81 has a large screen for both text and graphics displays, interactive graphics capability, and is programmable. If you approach the TI-81 as a computer, then you will better understand its many features and uses for the mathematics classroom.

In this manual we hope to provide examples and exercises that will explain how the TI-81 graphing calculator can be used to teach concepts and problem solving techniques in algebra. What will become apparent is that some (many) of the techniques and methods used in the pretechnology period are simply not appropriate when using powerful computational and graphing tools like the TI-81 graphing calculator. It is not the case that these graphing calculators will do the mathematics for the students or replace important skills students need to master. In fact, when using this new technology, students are asked to experiment, explore, and understand mathematical processes, concepts, and problem solving at a significantly deeper level than previously possible. Now a graph becomes the first step in understanding the problem and devising a plan for solution. Extending problems becomes a process of repeatedly altering the viewing rectangle and drawing multiple views of the graph.

A graphing calculator is a vehicle that allows students to make connections between the numerical, graphical, and symbolic representations of mathematical relationships. These connections can be made in a temporal setting that allows students to see the interrelationships between the various representations. Unlike many dedicated computer programs, the TI-81 can make these connections between the multiple representations of a relationship without lengthy delays for loading different software programs. The interaction can be customized by the students to suit their particular needs and modes of learning. The TI-81 allows students to tap the power of visualization to help them understand mathematical processes and problem solving strategies in ways not possible without a powerful graphing utility.

1

Numerical Operations and Linear Functions

1.1 Selecting Modes of Operation on the TI-81

The first thing you notice about the TI-81 graphing calculator is the large screen (see Fig. 1.1). This screen is like an electronic note pad with 8 lines, each capable of showing 16 characters. Each character or command you enter from the keyboard is written on the **home screen**.

When you first turn your graphing calculator on, you will see the **home screen**. Your calculator will perform a wide variety of mathematical procedures when set in different **modes**. To begin, press the [MODE] key to activate the **MODE** menu.

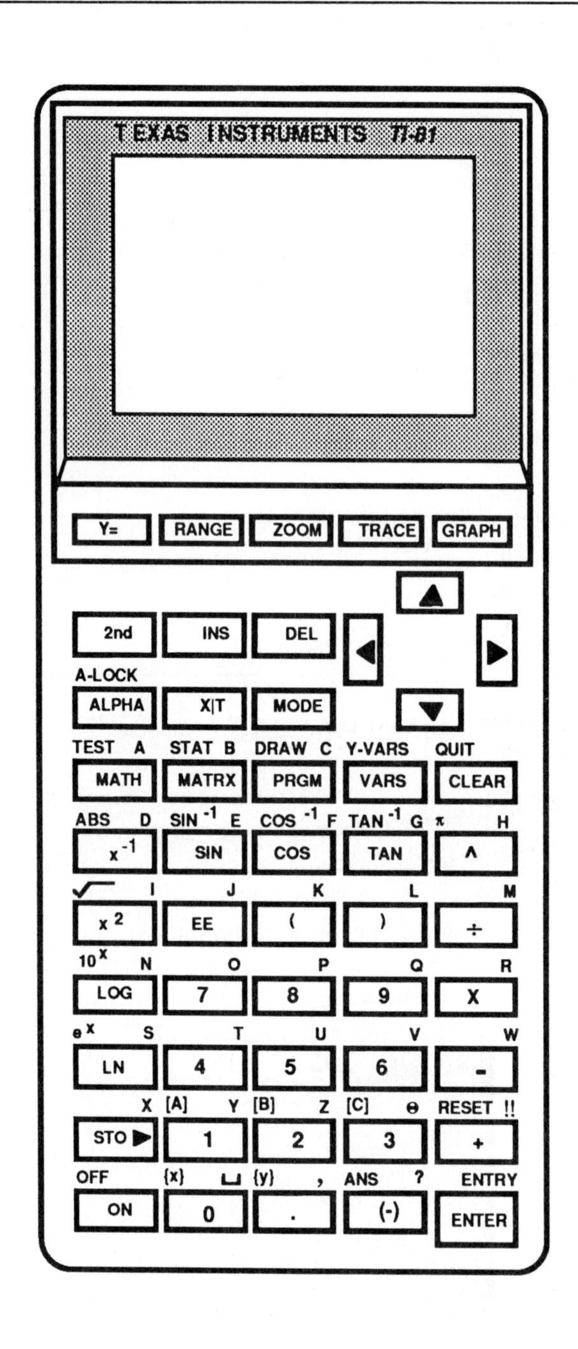

FIGURE 1.1
The TI-81 screen and keyboard

Figure 1.2 shows the screen with this menu active. Each row of settings has two or more choices. The leftmost choice in each row is the default setting. To select a row of choices, press the [▲] or [▼] cursor control arrow keys until the first entry in the row flashes. Use the [◄] and [►] cursor control keys to select one of the entries in a row. For example, in Figure 1.2, the angle measurement of degrees (**Deg**) has been highlighted (the word **Deg** flashes). Press the [ENTER] key to select the highlighted choice. In this case, the word **Rad** (for radians) will appear in normal text display after the choice of **Deg** is made by pressing the [ENTER] key.

If your calculator's **MODE** screen is not set as seen in Fig. 1.2, use the cursor control keys to select the various options, and press the [ENTER] key to activate each choice. At this point, place your calculator in **Deg** mode (angle measurement in degrees). Press the [CLEAR] key to return to the **home screen** when you are ready to continue.

To adjust the contrast of the screen, press the [2nd] key and hold down the [▼] arrow key to darken the screen or press the [2nd] key and hold down the [▲] arrow key to lighten the screen. Each level of contrast is numbered from 0 (lightest) to 9 (darkest). The number of the level appears in the upper right-hand corner of the viewing screen. If you cannot darken the screen enough to see the text and graphics at level 8 or 9, then you need to replace the batteries.

The [CLEAR] key will clear the text and numerical entries on the home screen. It is not necessary to push the [CLEAR] key before you begin a problem. In fact, many times you should **not** press the [CLEAR] key, because the information on the screen is useful when solving multistep problems. The screen will show up to four previous problems and answers at the same time. When doing multistep problems or using the results of one problem as the basis for the next problem, the previous results shown on the screen are very useful bits of information. Get out of the habit of pressing the [CLEAR] key each time you begin a new calculation. Just start entering the values and operations.

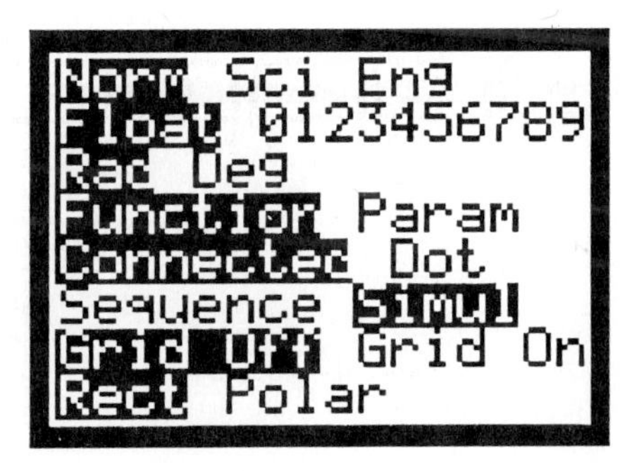

FIGURE 1.2
[MODE] screen

Remember, the ENTER key makes things happen! This key is like the RETURN key or the ENTER key on the computer. The key combination 2nd QUIT will also return operation to the **home screen.** This combination should be used to return to the **home screen** in cases where the CLEAR key will, in fact, clear part of the display, such as the Y= menu.

1.2 Exploring Numerical Calculations and Order of Operations

The TI-81 is a scientific calculator capable of all the numerical calculations that a "standard" scientific calculator can perform. The significant difference with the TI-81 is that all of the numerical entries, operational symbols, grouping symbols, and the answer to the calculation are visible on the screen at the same time. Since the screen has eight lines of display, multiple problems and answers can be seen at the same time. This multiple line display is one of the most powerful features of this calculator for teaching, learning, and problem solving. Let's investigate **algebraic order of operations** using some numerical problems.

EXPLORATION 1: Calculate the numerical value of $2 + 3 \times 4$.

Enter the numerical values and operations symbols in the same order they appear in the problem statement and press ENTER. The screen of the calculator should look like Fig. 1.3.

The answer to this problem is 14, not 20. The TI-81 uses a system of **algebraic hierarchy** to determine the order of operations. In the **algebraic order of operations,**

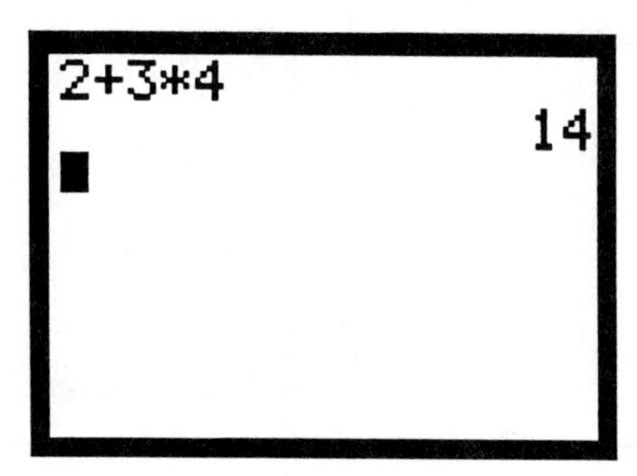

FIGURE 1.3

multiplication or division is done before addition or subtraction, from left to right. The calculator multiplied 3×4 first and then added 2 to get 14. ◊

EXPLORATION 2: Insert parentheses in the phrase $2 + 3 \times 4$ so the result is 20.

In order to get a result of 20, we must tell the calculator to add 2 + 3 before multiplying by 4. In the algebraic order of operations, quantities in parentheses are done before any other operations. If we group 2 + 3 in parentheses, the calculator will do this operation first. Without clearing the text screen enter the phrase using parentheses to group 2 + 3 (see Fig. 1.4). ◊

EXPLORATION 3: Compute the value of $\frac{12+6}{3-9}$.

This is a division problem. The fraction bar acts as a grouping symbol. In order to get the correct solution, the operations in the numerator and denominator must be done before the division. Figure 1.5 shows this problem entered on the screen. The entire numerator and denominator are grouped with parentheses. ◊

EXPLORATION 4: Enter the same sequence of key strokes for the problem in Exploration 3, but without any parentheses. What mathematical phrase is being evaluated?

Figure 1.6 shows the screen with the problem entered. The algebraic order of operations causes the division of 6 by 3 before the addition or subtraction is done. Once the division is done, 12 is added to the quotient and then 9 is subtracted, giving a result of 5. The actual mathematical phrase evaluated by this set of keystrokes is $12 + \frac{6}{3} - 9 = 5$. ◊

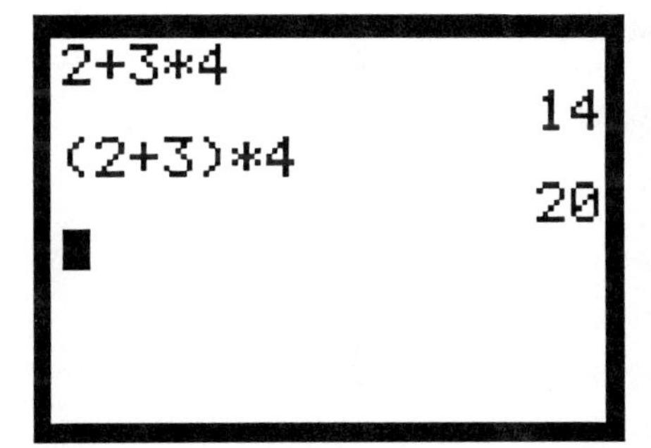

FIGURE 1.4

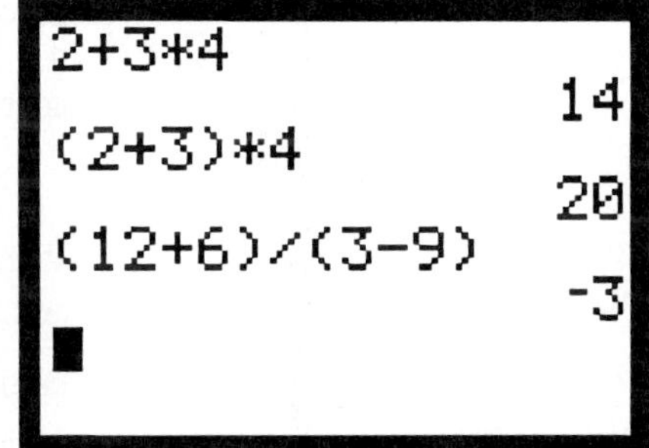

FIGURE 1.5

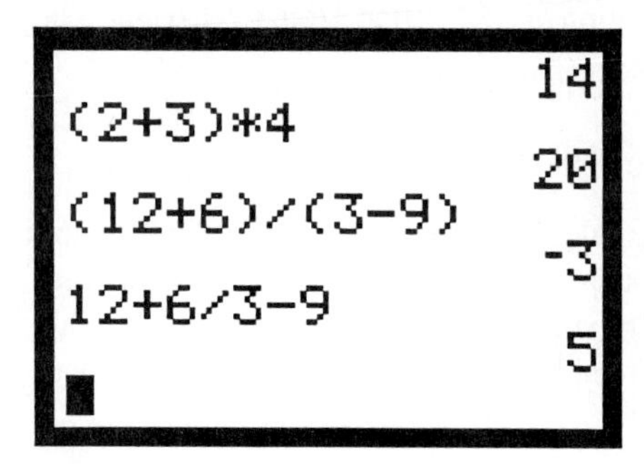

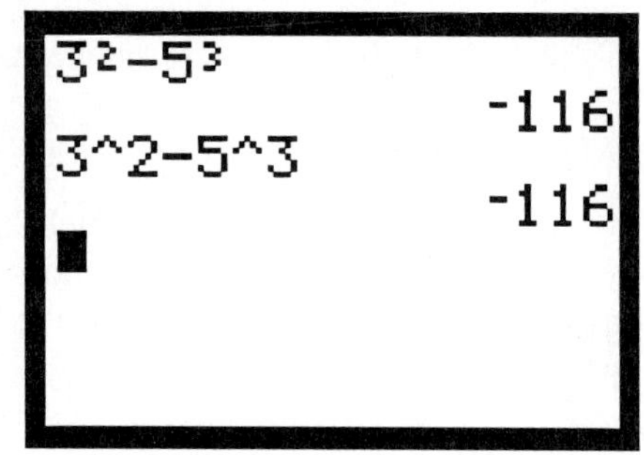

FIGURE 1.6

FIGURE 1.7

EXPLORATION 5: Calculate the value of $3^2 - 5^3$.

Exponentiation is done before subtraction in the order of operations. In this problem, the value 3 will be raised to the second power (or "squared") and the 5 will be raised to the third power (or "cubed") before the subtraction is completed. Figure 1.7 shows the screen with this problem entered in two different ways.

The TI-81 has dedicated keys to use when raising a number to the second or third power. The first entry in Fig. 1.7 shows the problem entered using the [X^2] key and the **[3: ^{3}]** menu option from the [MATH] menu. The more general way to raise a base to a power is to use the [^] key, as in the second entry in Fig. 1.7. As you can see, both results are identical. ◊

IMPLIED MULTIPLICATION, OR JUXTAPOSITION

Juxtaposition means positioning two symbols side by side. In mathematical symbolization, juxtaposition is used to imply multiplication. In the expression "5m" the juxtaposition of the 5 and the variable symbol m indicates "five times m." The juxtaposition of a number and a grouping symbol also implies multiplication. For example, the mathematical phrase "8(3 + 5)" means "eight times the quantity of three plus five." The graphing calculator understands juxtaposition as implied multiplication.

EXPLORATION 6: Evaluate the mathematical phrases $8 \times (3 + 5)$, $8(3 + 5)$, and $(3 + 5)8$ using your graphing calculator.

Figure 1.8 shows the three phrases entered on the graphing calculator; all phrases give the same result.

The second phrase, $8(3 + 5)$, is an example of the use of the **distributive property of multiplication over addition** with the multiplicative factor on the left. This is often

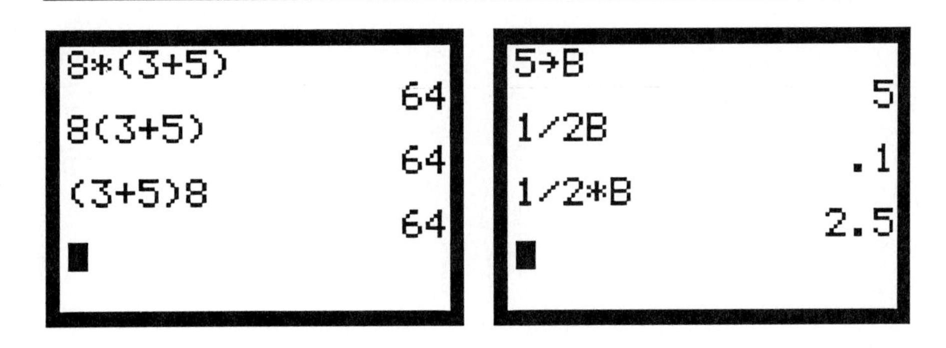

FIGURE 1.8

FIGURE 1.9

called the **left-hand distributive property**. The third phrase is an example of the **right-hand distributive property**. The graphing calculator understands these entries and gives the correct result for the implied multiplication. While juxtaposition is a way of indicating multiplication on a graphing calculator, it occupies a position in the order of operations above multiplication. This means that implied multiplication will be done before multiplication using the [×] key. ◊

EXPLORATION 7: The alphabetic characters on the TI-81's keyboard (usually accessed by using the [ALPHA] shift key) are the primary memory storage locations used by the calculator. Storing numerical values in these memory locations is done by using the [STO ▶] key. When this key is pressed, an arrow (→) appears on the screen indicating that the numerical value is placed in the memory location. For convenience, the [ALPHA] shift is automatically activated when the [STO ▶] key is pressed so the next character typed will be an alphabetic character. For example, the following keying sequence will store the value 5 in memory location **B** (see Fig. 1.9):

[5] [STO ▶] [B] [ENTER]

Estimate the value of the following phrases before you calculate the value:

1/2B and **1/2*B**

If $B = 5$, then by standard order of operations, 1/2B = 2.5 and $1/2*B = 2.5$. If implied multiplication is executed before standard multiplication or division, then $1/2B = .1$. Figure 1.9 shows both phrases evaluated. Implied multiplication is done before standard multiplication (*) or division (/). ◊

EXPLORATION 8: Calculate the value of $\frac{48}{3(6+2)}$.

Figure 1.10 shows the phrase entered in three different ways. In the first phrase, juxtaposition is used to indicate the multiplication of 3 and the quantity (6 + 2). After the quantity in parentheses was calculated (6 + 2 = 8), the multiplication of 3 and

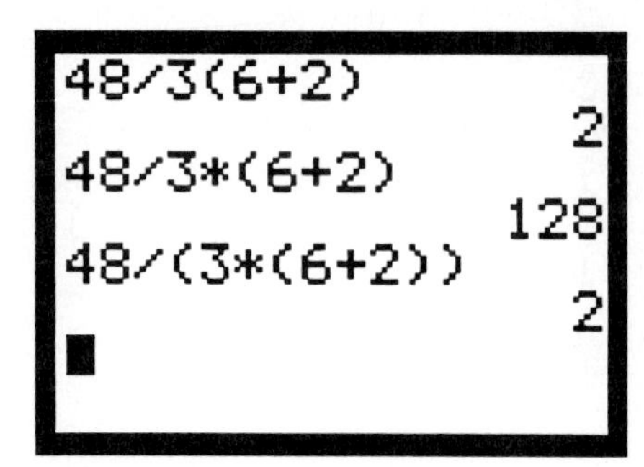

FIGURE 1.10

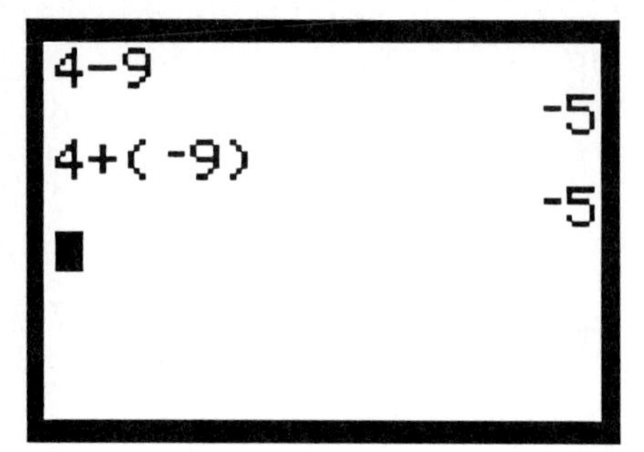

FIGURE 1.11

(6 + 2) was done to get 24. Then 48 was divided by 24 to get the result of 2. But, by the rules of algebraic order of operations, the division of 48 by 3 should have been executed before the multiplication of 3 and (6 + 2), since multiplication and division are done in order from left to right.

In the second phrase, the multiplication symbol is used to indicate the multiplication of 3 and (6 + 2). In this phrase, 48 is divided by 3 to get 16. Then 16 is multiplied by 8 (the sum of (6 + 2)) to get 128. This entry follows the normal algebraic order of operations. The mathematical phrase evaluated by this entry is $\frac{48}{3}(6+2)$ and not the original phrase.

The third entry shown in Fig. 1.10 shows the original phrase evaluated correctly using the multiplication symbol instead of juxtaposition. Notice that the entire denominator of the original phrase is enclosed in parentheses. Remember, a fraction bar is a grouping symbol and is represented on the graphing calculator by grouping the appropriate terms in parentheses. ◊

NEGATIVE NUMBERS AND SUBTRACTION

When doing mathematics using positive and negative numbers, we must make a careful distinction between the binary operation of subtraction and the sign of a negative number. For example, the term "-5" indicates "the opposite of five." The phrase "4 – 9" means "nine subtracted from four." When dealing with signed numbers and subtraction, it is often helpful to think of subtraction as the **addition of the opposite of a number.** The phrase "4 – 9" could be rewritten as "4 + (-9)," which is read "four plus the opposite of nine." Figure 1.11 shows the screen with both these forms entered.

Notice the difference between the screen symbols for subtraction and the opposite of a number. The symbol for the opposite of a number (the [(–)] key) is shorter and raised slightly compared with the symbol for subtraction (the [—] key).

REVIEW OF ALGEBRAIC ORDER OF OPERATIONS FOR THE TI-81 GRAPHING CALCULATOR

Parentheses	All quantities enclosed in parentheses are done according to the rules of order of operations.
Exponentiation	Any values raised to powers are calculated.
Implied multiplication	Multiplication implied by juxtaposition of two or more symbols; done from left to right.
Multiplication or division	Multiplication and division operations (indicated by the [×] or [÷]) are done in order from left to right.
Addition or subtraction	Addition and subtraction operations are done in order from left to right.

EXPLORATION 9: Calculate the value of $-12.5 - 3.2^2 + 4.5^2$.

Use the [(–)] key to enter "the opposite of 12.5" and use the [–] key to indicate the subtraction. Figure 1.12 shows the screen with the results.

Problem Extension

Enter the same problem from Exploration 9, but change the subtraction to the addition of the opposite. How should the squared quantities be handled? Figure 1.13 shows two versions of the problem. Which one is correct? Why? ◊

Standard algebraic order of operations is the same except that implied multiplication is treated as standard multiplication.

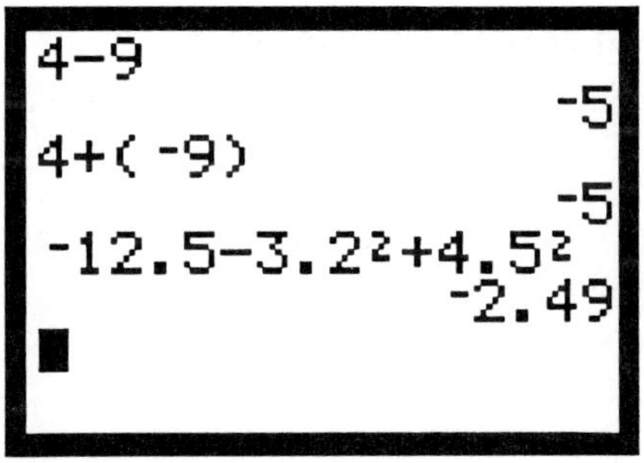

```
4-9
                -5
4+(-9)
                -5
-12.5-3.2²+4.5²
             -2.49
```

FIGURE 1.12

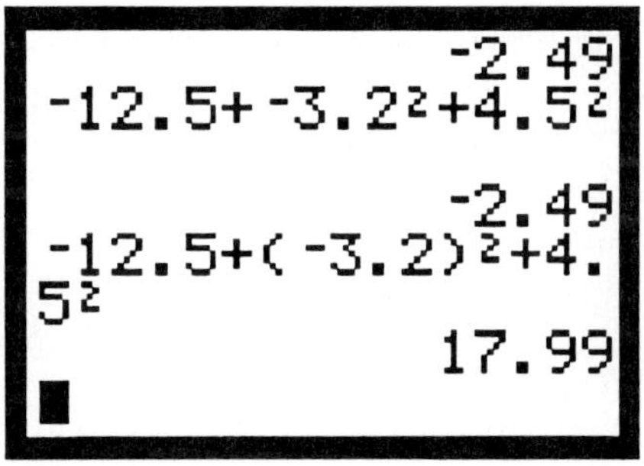

```
             -2.49
-12.5+-3.2²+4.5²
             -2.49
-12.5+(-3.2)²+4.
5²
             17.99
```

FIGURE 1.13

LAST ENTRY RECALL AND EDITING

The TI-81 has a special constant memory function, which will recall the last phrase entered into the calculator, even if the machine has been turned off. Like other computers, the TI-81 has editing features that allow the user to insert and delete characters in commands and program statements. These two features combine to form a powerful numerical problem solving process applicable to many types of problem situations.

EXPLORATION 10: The third power of a number is equal to 10; find the number.

Enter a first guess on the calculator and raise that number to the third power. Let's try 5. Is the result 10? (See Fig. 1.14.) Do we need to make the next guess higher or lower? Press the [▲] arrow key, and the TI-81 will recall the last entered phrase and place the cursor at the end of the phrase for editing. Figure 1.14 also shows the first phrase recalled and ready for editing.

Use the [◀] and [▶] arrow keys to move the cursor in position over the 5 and overwrite with a 4 by simply typing a 4 to change the problem (see Fig. 1.15). (The cursor moves over the next character and flashes.)

Press [ENTER] to evaluate the new phrase (see Fig. 1.16). Is the new result 10? How should we change the number to get an answer closer to 10? Use the [▲] arrow key to recall the last entered phrase again, and change the 4 to a 3 (the base value of the exponent). Continue this process and find what two whole numbers the answer lies between (2 and 3; see Fig. 1.17).

Now let's refine our guesses to get a much more accurate answer to our problem. Recall the last entry (2^3). Use the arrow keys to move the cursor over the exponentiation symbol (^) and press the [INS] key to activate the insert mode. When you press the [INS] key, the cursor changes from a black rectangle to an underline and flashes at its present position (see Fig. 1.18).

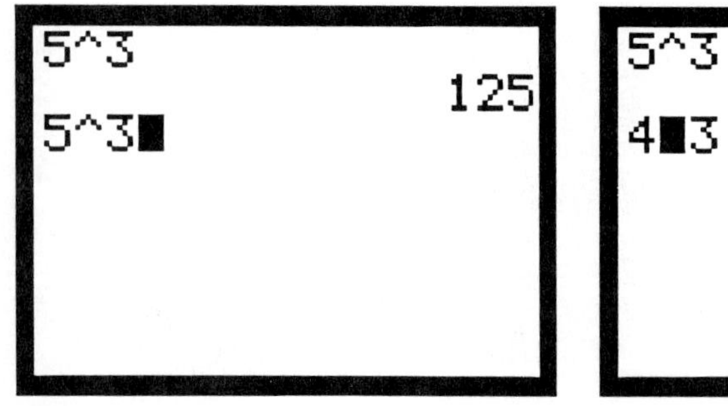

FIGURE 1.14

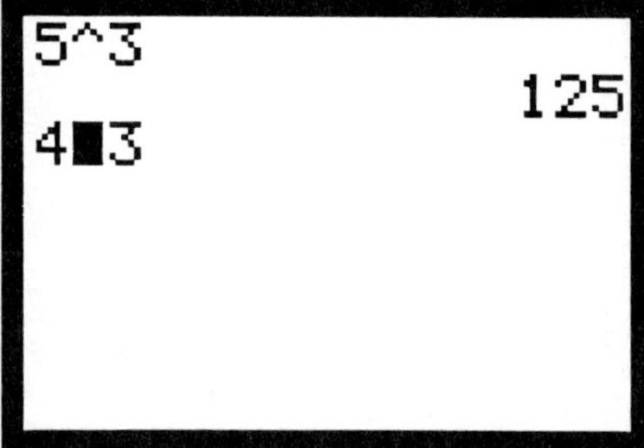

FIGURE 1.15

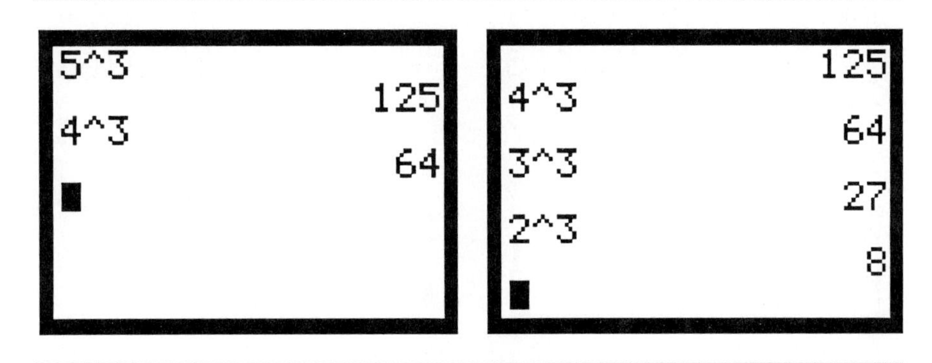

FIGURE 1.16 FIGURE 1.17

Insert the characters ".5" after the 2 to make the phrase "2.5^3." Press the ENTER key to evaluate the new phrase (see Fig. 1.19).

Is the new result greater or less than 10? How should we change the base value for our next guess? Recall the last phrase and change the base to 2.4 by overwriting a 4. If this is not close enough, continue the process until the result is less than 10. Figure 1.20 shows the screen for the series of base values 2.4, 2.3, 2.2, and 2.1. Unwanted characters may also be deleted from the phrase any time by simply placing the cursor on the character to be deleted and pressing the DEL key.

Repeat this insert-and-edit process for the next decimal place in the base. If you continue the process, eventually you will get a screen similar to Figure 1.21 showing the number raised to the third power that gives a result of 10.

Problem Extension

What we solved in the preceding is the equation $x^3 = 10$. Another way to view this is that we found $\sqrt[3]{10}$. Calculate $\sqrt[3]{10}$ directly using choice 4 from the MATH menu,

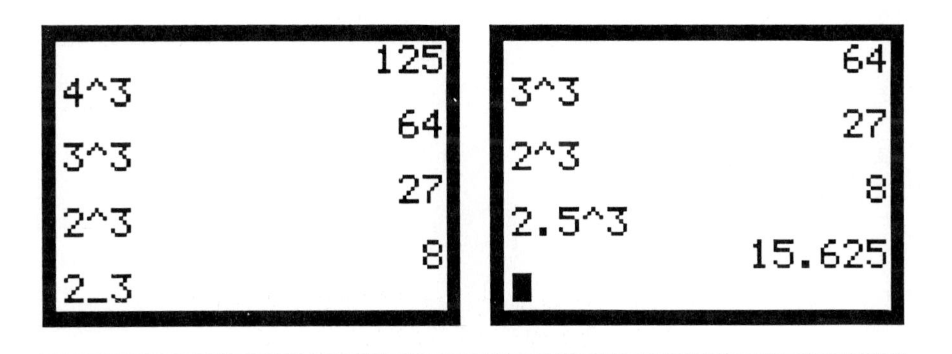

FIGURE 1.18 FIGURE 1.19

```
              15.625
2.3^3
              12.167
2.2^3
              10.648
2.1^3
               9.261
```

FIGURE 1.20

```
         9.999999303
2.15443465^3
         9.999999443
2.15443468^3
          9.99999986
2.15443469^3
                  10
```

FIGURE 1.21

and compare it with the value we found by the guess-and-check method. Are they similar? But, is $\sqrt[3]{10}$ a rational or irrational number? How can the calculator give an exact value for $\sqrt[3]{10}$ if it is irrational? This discussion raises important mathematical questions which students need to address. ◊

Problems

Use your TI-81 graphing calculator to evaluate the following mathematical phrases.

1. $12.8\,(7.9 - 18.75)$

2. $-137.5 - 15.9$

3. $35^2 - 25^2$

4. $187.56 + 392.33 - 29.99$

5. $\dfrac{0.0725 + 1.23}{6.95 - 9.7}$

6. $14 - \dfrac{12}{9} + \dfrac{7}{8}$

7. $4\,(7 - 8) + (-3)(29 - 34)$

8. $\dfrac{4^6}{16^3}$

9. $(9.3 + 16.5)8.5$

10. $-4^4 - 5^4$

Arrange each set of values in order from smallest to largest.

11. $3.6^4, 3.6^5, 3.6^1, 3.6^{-2}$

12. $.5^2, .5^3, .5^0, .5^{-1}$

13. $\dfrac{2222}{22222}, \dfrac{222}{2222}, \dfrac{22}{222}, \dfrac{2}{22}$

14. $\left(\dfrac{3}{2}\right)^3, \left(\dfrac{3}{2}\right)^2, \left(\dfrac{2}{3}\right)^2, \left(\dfrac{2}{3}\right)^3$

Write the mathematical phrase represented by the following screen displays.

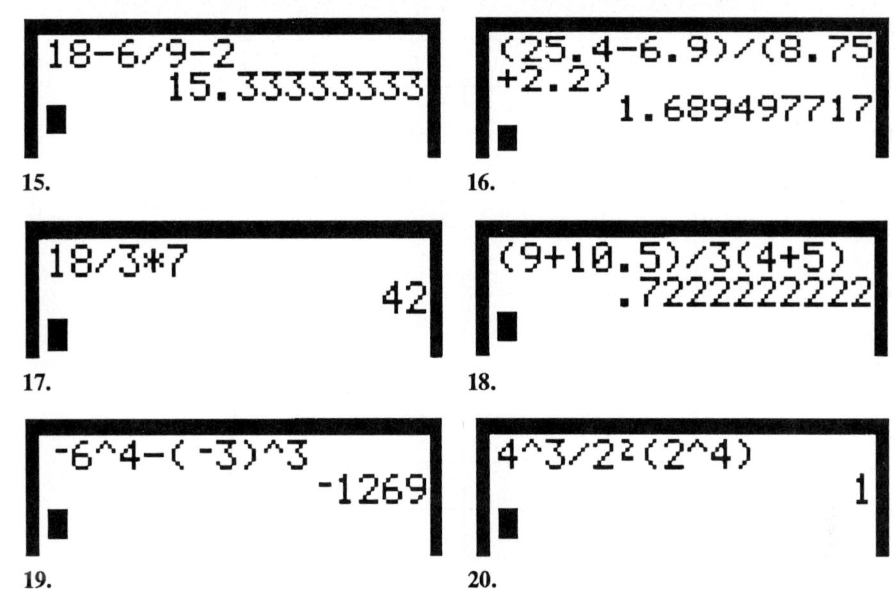

List the order in which the following mathematical phrases will be evaluated by the TI-81 graphing calculator.

21. $5^2\left(8+\frac{6}{7}\right)$

22. $\dfrac{6^2-3^2}{9-\frac{5}{7}}$

Complete the following tables.

23. The second number is three times the square of the first number plus five.

First number	Second number
0	
1	
2	
3	
4	
5	
	152
15	

24. The second number is 15 less than two times the second power of the first number.

First number	Second number
-5	35
-2	
-1	
0	
1	
2	
5	
	147
12	
N	

FOR ADVANCED ALGEBRA STUDENTS

Using the last entry recall and editing features of the TI-81, find a value for the number that satisfies the following conditions.

25. The number raised to the third power is 20.

26. The number raised to the fifth power is -0.1.

27. The number squared less five times the number is -6.

28. Four times the number to the third power minus the number squared is 12.

29. Standing in the front of a boat, the area of water a fisherman can cast to is a circular region with the fisherman at the center (ignore the area of the boat). Make a table that shows the area a fisherman can cast to for various length casts from 0 to 50 meters.

30. Expand the table you make in problem 29 to show the effect of increasing the length of each cast by 10% on the area of water the fisherman can cast to. For example, if the original cast was 10 meters, the new cast would be 10 + (.10) 10 meters, or 11 meters.

31. The Biawa Tackle Co. claims that its new "Far Cast" reel will cast 10% farther than ordinary reels of the same kind. This increase in the length of each cast, so the company claims, gives the fisherman a 20% increase in the area of water that the fisherman can effectively fish. Using your table from problem 30, determine if the Biawa Tackle Co.'s advertising claims are correct or misleading. What is the actual increase in fishable area gained by a 10% increase in cast length?

32. The Third National Bank pays 7.5% annual interest on savings account deposits. If the bank compounds interest monthly, make a table showing the amount of money on deposit at the end of 1, 2, 4, 5, 8, and 10 years. Does the amount earned double from 2 years to 4 years? From 5 years to 10 years?

33. Using the Third National Bank's rate of 7.5% and monthly compounding period (from problem 32), how long will it take for \$1200.00 to triple in value?

34. If you had deposited \$5000.00 in your account at the Third National Bank, how long would it take to triple at 7.5% interest compounded monthly?

35. For a deposit of \$1200.00 at 7.5% interest, make a table comparing the amount earned in 5 years for the following compounding periods: yearly, monthly, weekly, daily, hourly, every minute, and every second.

1.3 Exploring the Viewing Rectangle, the Coordinate Plane, Points, and Lines

The effectiveness of any graphing utility depends on understanding the viewing window of the utility and its relationship to the coordinate plane and the scale of the view. The coordinate plane in which we will draw graphs extends in all directions to infinity. Since we are working with a finite machine, we must define a portion of the coordinate plane that will appear within the viewing window of the graphing calculator. This window is known as the **viewing rectangle** or **viewing window.** To set the dimensions of the viewing rectangle, press the [RANGE] key to display the **Range** screen. Figure 1.22 shows the **Range** screen with the TI-81's **Standard,** or default, viewing rectangle entered.

The values for **Xmin** and **Xmax** refer to the horizontal dimension of the viewing rectangle. An alternate way to interpret these values is to say that **Xmin** is the **left edge** of the screen and **Xmax** is the **right edge** of the screen. **Xscl** indicates what each tick mark on the horizontal axis represents. For example, **Xscl = 1** indicates that each tick mark represents 1 unit. Alternate interpretations are that the scale gives the

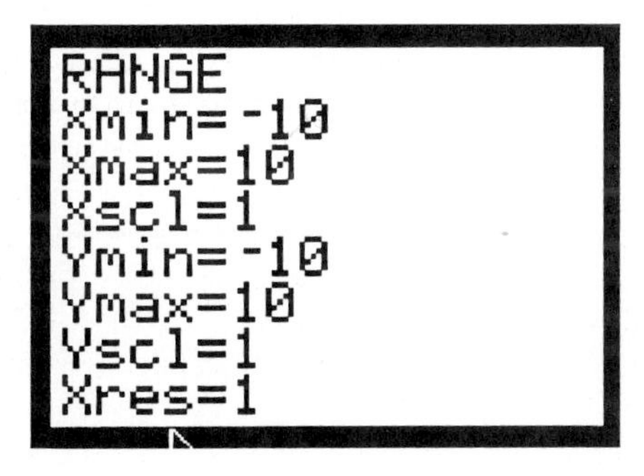

FIGURE 1.22
Range screen

distance between each consecutive tick mark, or the distance from the origin to the first tick mark. **Ymin, Ymax,** and **Yscl** refer to similar measurements in the vertical direction. **Ymin** is the bottom of the screen, **Ymax** is the top of the screen, and **Yscl** is the distance between tick marks in the vertical direction. Figure 1.23 shows the graphics screen representing the viewing rectangle set by the values shown in Fig. 1.22.

The values for **Xscl** and **Yscl** are a matter of personal choice. While some scale values are not appropriate for a given viewing rectangle (such as **Xscl** of 1 for **Xmin** of -100 and **Xmax** of 100), often there are several values that will work well. For example, in a **Standard** viewing rectangle of [-10, 10] by [-10, 10], **Xscl** and **Yscl** values of 1, 2, or 5 would all be appropriate. If you set the scale too small for the viewing rectangle, one or both of the axes may appear as a double line. This is caused by the tick marks being too dense (close together) on the axis, causing them to appear as a solid line. Scale values may be set differently on the horizontal and vertical axes to accurately reflect the size of the viewing rectangle being used. If you want to eliminate the tick marks, enter a 0 for **Xscl** and/or **Yscl.**

EXPLORING THE VIEWING RECTANGLE

EXPLORATION 1: Set your calculator in the standard viewing rectangle by entering the values shown in Fig. 1.22. Remember to use the [(–)] key to indicate the "opposite of 10" and not the [—] key for subtraction. After each value is entered, press either the [ENTER] key or the [▼] arrow key to move to the next line. If your cursor is in the first position of any **Range** entry line, the previously entered value will be cleared when the first new entry value is keyed. If the cursor is in any other position in a line, only that character will be altered.

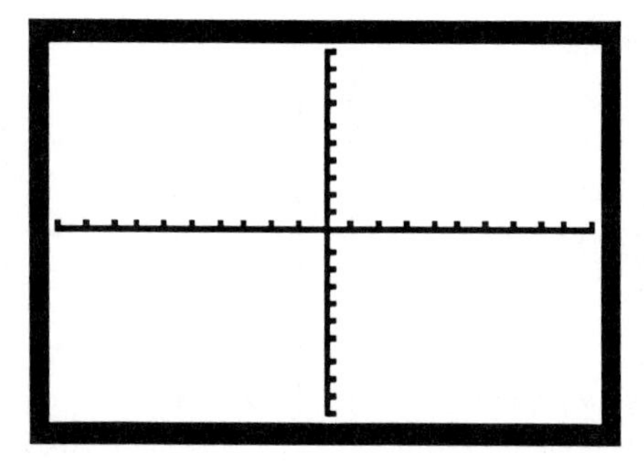

FIGURE 1.23
Graphics screen

Press the [Y=] key to activate the menu of functions. Using the [▲] or [▼] arrow keys and the [CLEAR] key, clear all functions from the menu. Press the [GRAPH] key and the blank graphics screen will appear as shown in Fig. 1.23. Press any of the four arrow keys and a small cross with a flashing center will appear near the center of the screen. This is known as a **screen cursor.** At the bottom of the screen the x- and y-coordinates of this cursor are shown (see Fig. 1.24).

The screen cursor is controlled by the arrow keys. If you press an arrow key, the cursor will move one pixel (a rectangular block of light on the screen) in that direction, and the new coordinates of that point will be displayed. If you press and hold an arrow key, the screen cursor will "run" in that direction. Run the screen cursor to the edges of the viewing rectangle. Are these values what you expected? Do they exactly match the values you entered on the [RANGE] screen? Figure 1.25 shows the screen cursor in the upper left corner of the viewing rectangle and the coordinates **X = -10** and **Y = 10** at the bottom of the screen.

A very important feature of the graphics screen is the numeric distance between each pixel on the screen in the horizontal and vertical directions. As you step around the graphics screen, note the distance between each pixel horizontally and vertically. You may do this numerically by subtracting the coordinates of one position from the coordinates of the next pixel to the right and above. For example, set your screen cursor on the position **X = 2.4210526** and **Y = 4.6031746** as shown in Fig. 1.26. Move one pixel to the right and one pixel up and find the difference horizontally and vertically.

In this case with the default range of [-10, 10] by [-10, 10], 2.6315789 – 2.4210526 = .2105263 and 4.9206349 – 4.6031746 = .3174603. Try any other pair of coordinates. Do you get the same values for the difference? In fact, each diagonal pair of pixels differs by the same amount in the horizontal and vertical direction (given slight round-off errors).

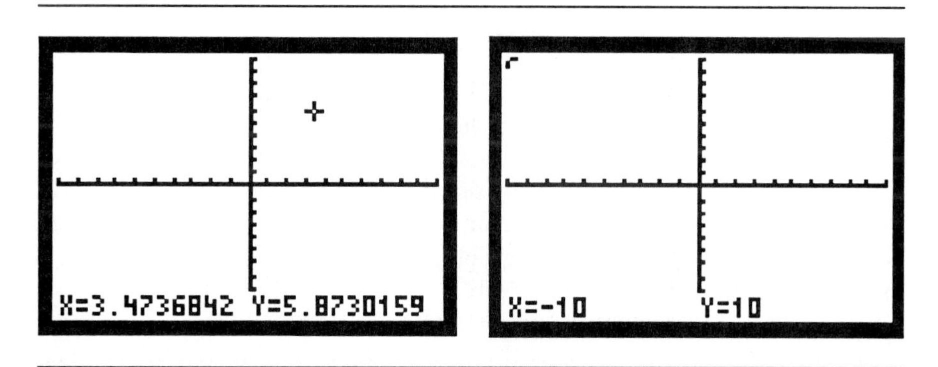

FIGURE 1.24
Screen cursor

FIGURE 1.25
Cursor at (-10, 10)

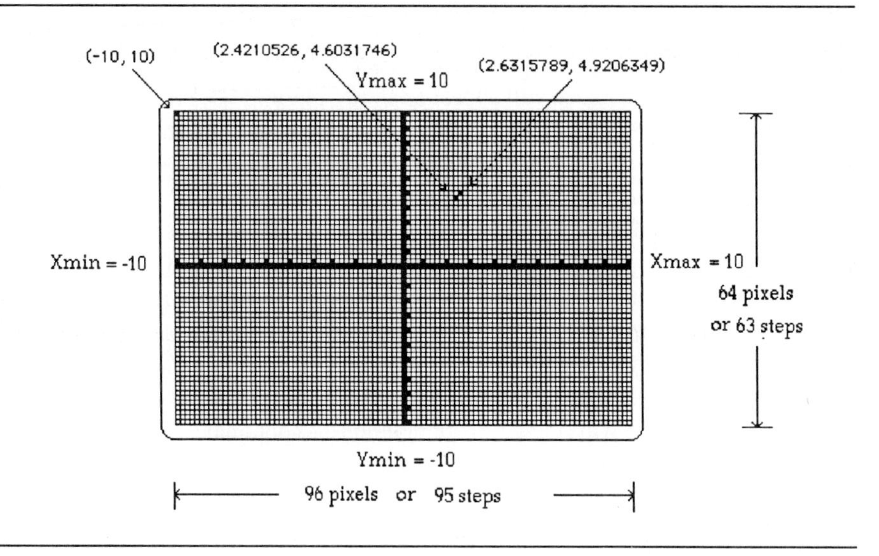

FIGURE 1.26
The graphics screen made up of 96 × 64 pixels

The size of the step between each pixel is a function of the range you set and the number of pixels on the graphics screen. The TI-81 graphics screen consists of 96 vertical columns and 64 horizontal rows of pixels (see Fig. 1.26). In the horizontal direction, if the first pixel on the left represents -10 and the last pixel on the right represents 10, as in the previous exploration, then how much should each equal step be to make the values work out correctly? If you were walking from the first pixel to the 96th pixel, how many steps would you take? The answer is 95, since there is always one less step than stepping stones. The numerical distance across the screen is $10 - (-10) = 20$ units, so the size of each step should be $\frac{20}{95} = .2105263158$ units. Similarly in the vertical direction, there are 64 pixels and 63 steps. If the numerical distance across the screen is $10 - (-10) = 20$, then each step is $\frac{20}{63} = .3174603175$ units. Do these two values look familiar? They are the same as the values we calculated by investigating successive points on the graphics screen.

The ratio of the length of each pixel step in the vertical direction to the length of each pixel step in the horizontal direction is called the **aspect ratio** of the graphics screen. For the default viewing rectangle of [-10, 10] by [-10, 10], the aspect ratio is $.3174603175/.2105263158 = 1.507936508$. ◊

EXPLORATION 2: Change the RANGE screen of your calculator to the settings shown in Fig. 1.27. Figure 1.28 shows the resulting graphics screen.

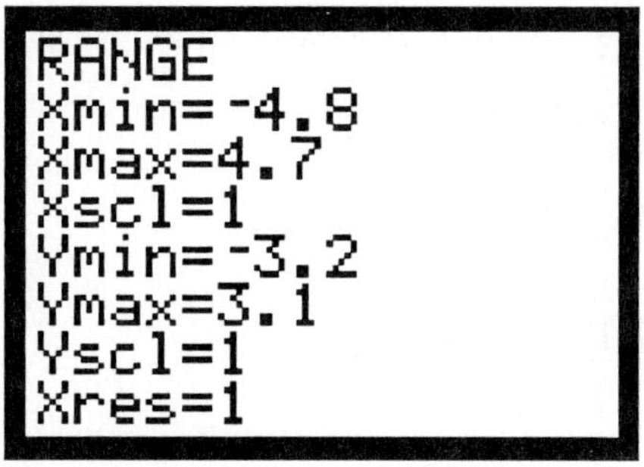

RANGE
Xmin=-4.8
Xmax=4.7
Xscl=1
Ymin=-3.2
Ymax=3.1
Yscl=1
Xres=1

FIGURE 1.27
Range screen

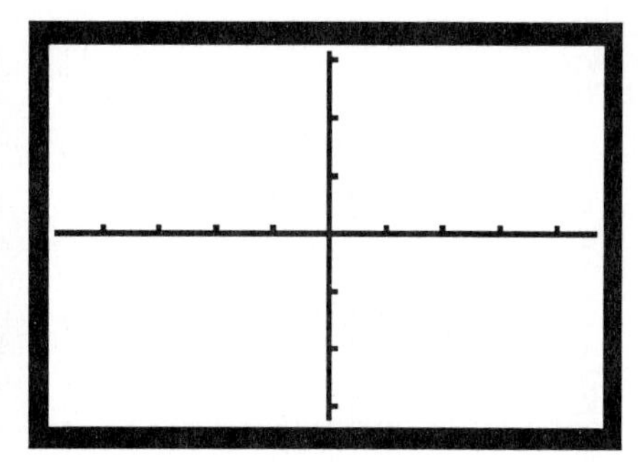

FIGURE 1.28
Graphics screen

Use the screen cursor to walk around this viewing window. Do you notice anything different about the values of the various points on the screen or the steps between each pixel? Calculate the numerical distance across the screen in both the horizontal and vertical directions (4.7 – (-4.8) = 9.5 and 3.1 – (-3.2) = 6.3). Divide each of these values by the appropriate number of steps (i.e. 95 or 63). What is the result? (9.5/95 = .1 and 6.3/63 = .1.) Notice that the step size in the horizontal direction equals the step size in the vertical direction. What is the aspect ratio of this graphics screen? Does the screen cursor step exactly on the integer values between -4 and 4 in the horizontal and -3 and 3 in the vertical direction? ◊

CHANGING THE SIZE OF THE VIEWING RECTANGLE

There are several different ways to change the size of the viewing rectangle. In the previous examples, values on the RANGE screen were set individually. This is the fundamental method of setting the viewing rectangle. The viewing rectangle may also be set by using built-in key functions or through programming.

EXPLORATION 3: The ZOOM menu contains four options that automatically set the viewing rectangle. Figure 1.29 shows the ZOOM menu scrolled downward to show options 2 through 8.

When selected, these options 5 through 8 will automatically set the viewing rectangle in different ways. Option 5, **5: Square,** alters the set viewing rectangle so that the aspect ratio is approximately 1. Figure 1.30 shows the RANGE screen after the **5: Square** option has been used on the **standard** viewing rectangle. Squaring the view-

```
ZOOM
2↑Zoom In
3:Zoom Out
4:Set Factors
5:Square
6:Standard
7:Trig
8:Integer
```

FIGURE 1.29
ZOOM menu

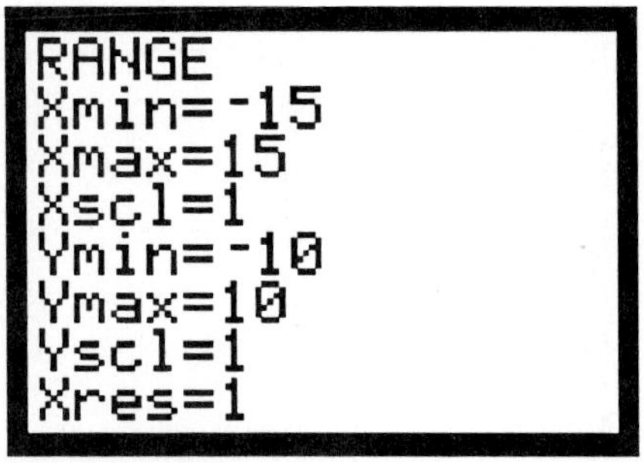

```
RANGE
Xmin=-15
Xmax=15
Xscl=1
Ymin=-10
Ymax=10
Yscl=1
Xres=1
```

FIGURE 1.30
Effect of 5: Square

ing rectangle produces visually correct representations of geometric shapes such as circles or perpendicular lines. Enter the following two linear equations on the Y= menu:

$$Y_1 = 2X - 3 \quad \text{and} \quad Y_2 = -.5X + 2.$$

(Press the Y= key to access the menu; use the X / T key to enter the variable **X.** Be sure to use the (–) key to enter -.5.)

The graphs of these two equations should be perpendicular lines crossing at the point (2, 1). Figure 1.31 shows the graphs of these two equations drawn by selecting option 6 from the ZOOM menu. Selecting option 6, **6: Standard,** will set the default viewing rectangle of [-10, 10] by [-10, 10] and graph the entered equation

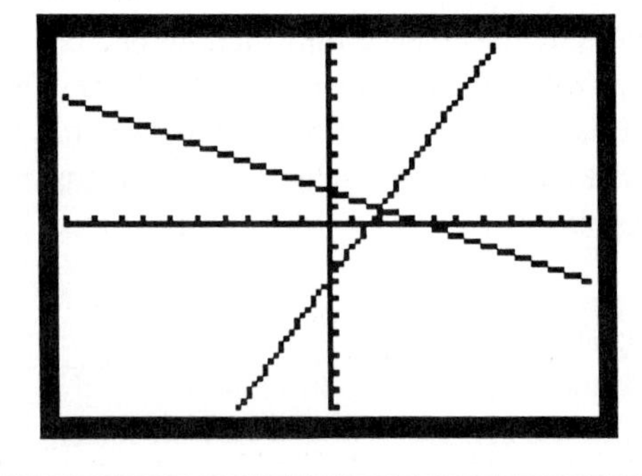

FIGURE 1.31
Standard view

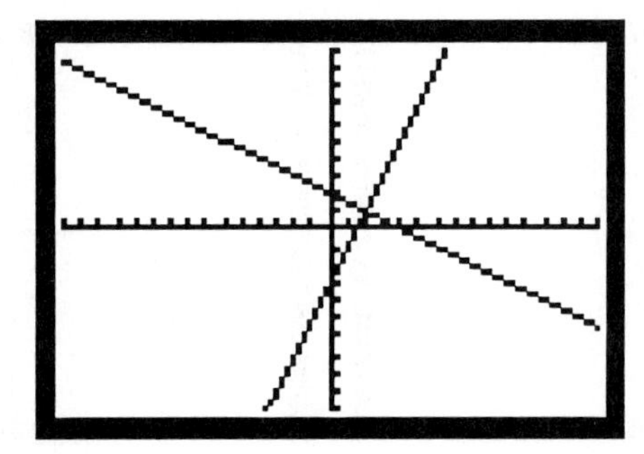

FIGURE 1.32
Squared view

immediately. Notice that the two lines **do not** appear perpendicular in this view. This happens when the aspect ratio of the screen is not equal to 1.

Figure 1.32 shows the same two equations graphed by selecting option 5 from the ZOOM menu, **5: Square**. Notice that the two lines appear to be perpendicular in this view. The aspect ratio is approximately 1 for this given range.[1]

Option 7, **7: Trig**, will set a viewing rectangle of [-6.28318531, 6.283185307] by [-3, 3] which is helpful for graphing trigonometric functions. Option 8, **8: Integer**, produces a viewing rectangle with each pixel representing an integer value in both the horizontal and vertical directions. These two options will be used in later examples.◊

USER DEFINED VIEWING RECTANGLES USING A PROGRAM

Another way to set viewing rectangles is through a program stored in the memory of the TI-81. Programs for setting a viewing rectangle can have various types of inputs to control the center of the window or the scaling factor in either the horizontal or vertical direction. For more information on variations of this idea, consult the fourth volume in this series, *Exploring Programming on the TI-81*. The following example lists a basic range setting program that gives viewing rectangles centered at the origin with aspect ratio of 1. Multiples of the basic screen can be obtained by entering a value other than 1 for the input "**RANGE FACTOR =**" in the program.

EXPLORATION 4: In Exploration 2 we investigated a viewing rectangle that had an aspect ratio of exactly 1 and steps of .1 between pixels. We also discovered that some

[1]Actual calculation of the aspect ratio of this screen based on the viewing rectangle [-15, 15] by [-10, 10] gives a value of 1.005291005. Visually this aspect ratio gives a correct looking picture; mathematically, however, this value does not produce a viewing rectangle in exactly square coordinates.

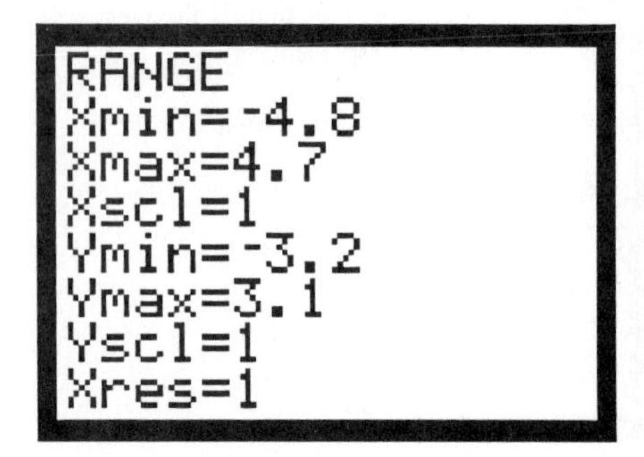

FIGURE 1.33

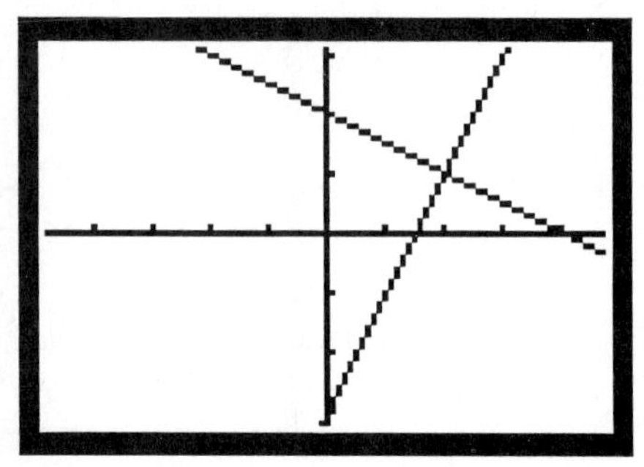

FIGURE 1.34

of the values that were represented on the screen were the integers between -4 and 4 in the horizontal direction and -3 and 3 in the vertical direction.

Figure 1.33 shows the RANGE screen for this **friendly** viewing rectangle. Figure 1.34 shows the previous two linear equations (from Exploration 3) graphed in this viewing rectangle. It is also true that multiples of these friendly range values produce viewing rectangles with an aspect ratio of 1 and "friendly" steps between pixels. For example, if each of the four basic range values is multiplied by 2, the resulting values, [-9.6, 9.4] by [-6.4, 6.2], produce a viewing rectangle with steps of .2 between each pixel, an aspect ratio of 1, and pixels that represent integer values.

Figure 1.35 shows the RANGE screen for these values, and Fig. 1.36 shows the resulting graph of the two linear equations. Notice that the lines still appear perpendicular in this "larger" window. Press any of the arrow keys and move the screen cursor around in this viewing rectangle to confirm the values represented in the window.

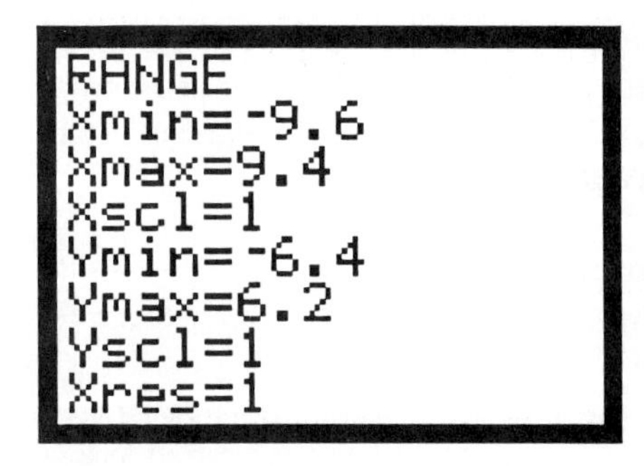

FIGURE 1.35

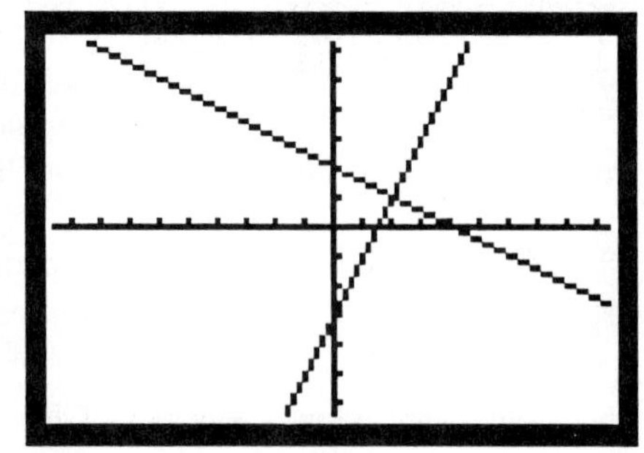

FIGURE 1.36

By exploring the graphics screens produced by multiplying the basic friendly values by various factors, you will discover the advantages of using these viewing rectangles. The aspect ratio is always 1. The steps between each pixel are $\frac{1}{10}$ of the multiplicative factor; multiplying by 1 gives steps of .1, by 2 gives steps of .2, and by 5 gives steps of .5. For certain factors, such as .5, 1, 2, 5, and 10, exact integer values are in the set of points represented in the viewing rectangle.

To write a program for the TI-81 that will automatically set the friendly viewing rectangle or multiples of the friendly viewing rectangle, press the [PRGM] key to access the **program storage** menu (see Fig. 1.37). Select the "**EDIT**" option and then select "**1: Prgm 1**" for program storage location #1. (If you already have a program stored in location #1, use any other empty location for this program.) A blank programming screen will appear with the cursor positioned in the **title line**, ready to enter the title of the program.

Notice that the cursor (**A**) is locked in the **Alpha** typing mode; when keys are pressed, the alphabetic characters above the keys will be typed. Type the name of the program as **RANGE**, and press [ENTER] (see Figure 1.38). ◊

Writing the Program

Once the program is named, enter the following code (listed on the left in the following table). The beginning of each line is denoted by a colon (:). Press [ENTER] at the end of each command line to go to the next line. As characters are entered on the screen of the TI-81, they are saved in the constant memory of the calculator; no special commands are needed to save your program. The program will remain in the memory of the calculator until you erase it or **Reset** the entire memory of the unit.

When the cursor is in any line of a program (except the title line), the [PRGM] key accesses a menu of programming commands including the commands **Disp** and **Input** located on the **I/O** submenu (see Fig. 1.39). Command words such as **Disp** or **Xmin** written in upper and lowercase type are accessed from menus or single key

FIGURE 1.37

FIGURE 1.38

USER DEFINED RANGE PROGRAM (CENTERED AT THE ORIGIN WITH ASPECT RATIO = 1)

Code	*Comments*
: Disp "RANGE FACTOR"	[prompt message]
: Input F	[input the range factor; store in F]
: -4.8F → Xmin	[multiply -4.8 by F and set Xmin]
: 4.7F → Xmax	[multiple 4.7 by F and set Xmax]
: F → Xscl	[set Xscl equal to input factor]
: -3.2F → Ymin	[multiply -3.2 by F and set Ymin]
: 3.1F → Ymax	[multiply 3.1 by F and set Ymax]
: F → Yscl	[set Yscl equal to input factor]

strokes. Words appearing in uppercase type only, such as "**RANGE FACTOR**," are typed one character at a time from the keyboard.

The assignment arrow (→) is accessed with the [STO ▶] key. Read this symbol as the phrase "is stored in" in command lines. For example, **: F → Xscl** is read as "The value stored in F is stored in Xscl." Remember that juxtaposition implies multiplication; the command **-4.8F** means "-4.8 × F." The [RANGE] variables are located on the **RNG** submenu of the **VARS** menu (variables), accessed by the [VARS] key (see Fig. 1.40).

Running the Program

To run the program, return to the **home screen** (key [2nd] [QUIT]). Press the [PRGM] key and select **1: Prgm 1 RANGE** by typing [1] (for the number of the program location) or by moving the cursor to the program location and pressing [ENTER]. The command **Prgm 1** is written on the **home screen**; press [ENTER] to

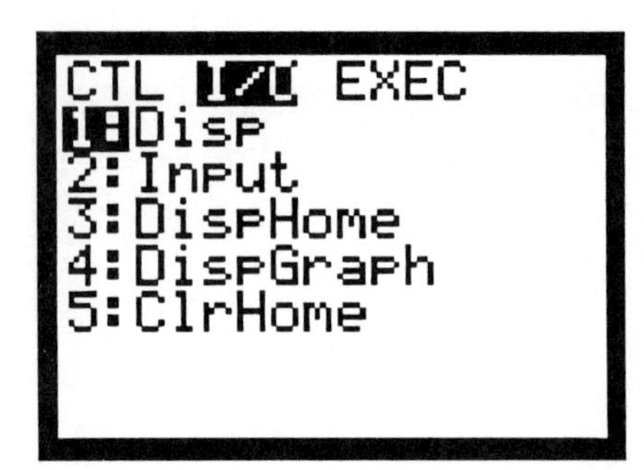

FIGURE 1.39
The I/O menu using the [PRGM] key

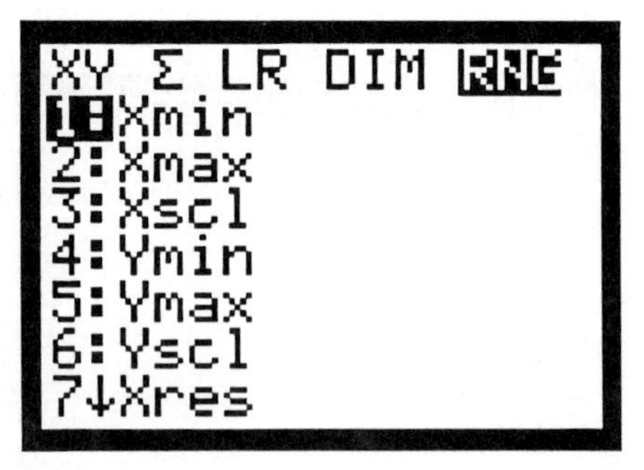

FIGURE 1.40
The RNG menu using the [VARS] key

execute the program. The program will prompt you for an input value; type 1 and press [ENTER] . The word "Done" will appear, indicating that the program has been executed. Examine the [RANGE] screen to see if the correct values have been set. Run the program again, and enter a factor of 2 and check the [RANGE] screen to see if the correct values are entered.

PLOTTING POINTS AND DRAWING LINES

The TI-81 can plot individual points or draw a specified line on the graphics screen. These functions can be done in two ways, numerically and interactively. The interactive functions are accessed by activating the **Draw** menu (key [2nd] [DRAW]) from the **graphics** screen, while the numerical functions are accessed by activating the [DRAW] menu from any other screen.

EXPLORATION 5: The Clason family is planning a trip from Marquette, Michigan to Chicago, Illinois, a distance of 500 miles, for a family visit. Brenda Clason is estimating the amount of time the trip will take for various average speeds. She knows that travel time is equal to the distance traveled divided by the average speed (rate), or $t = \frac{d}{r}$. She made the following chart of values:

Average speed (mph)	10	20	30	40	50	60	70	80	90
Travel time (hr)	50	25	16.67	12.5	10	8.33	7.14	6.25	5.55

Set the [RANGE] screen of the calculator based on the values in the table (see Fig. 1.41). *Note*: The graphics screen may be cleared by changing any of the [RANGE] settings, by changing or retyping any character on the [Y=] menu, or with the **1: ClrDraw** command from the [DRAW] menu.

Select the [DRAW] menu by keying [2nd] [DRAW] (see Fig. 1.42). Be sure you **do not** select the [DRAW] menu from the graphics screen. Select **3: Pt-On** (and the command will appear on the **home screen.** Enter the first pair of coordinates representing average speed and travel time, (10, 50). The two values are entered separated by a comma (key [ALPHA] [.]) and followed by a right parenthesis to close the phrase. Figure 1.43 shows the command entered on the **home screen.** Press [ENTER] to plot the point on the graphics screen (see Fig. 1.44).

To plot the next point, press the [CLEAR] key to return to the **home screen.** Press the [▲] arrow key to replay the last command, and edit the command using the [INS] and the [DEL] keys to change the values for the next point, (20, 25). Press [ENTER] to plot the next point. This process continues for each point in the table

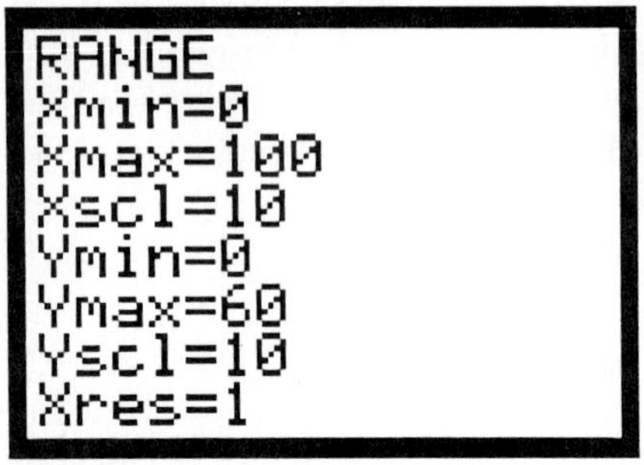

RANGE
Xmin=0
Xmax=100
Xscl=10
Ymin=0
Ymax=60
Yscl=10
Xres=1

FIGURE 1.41

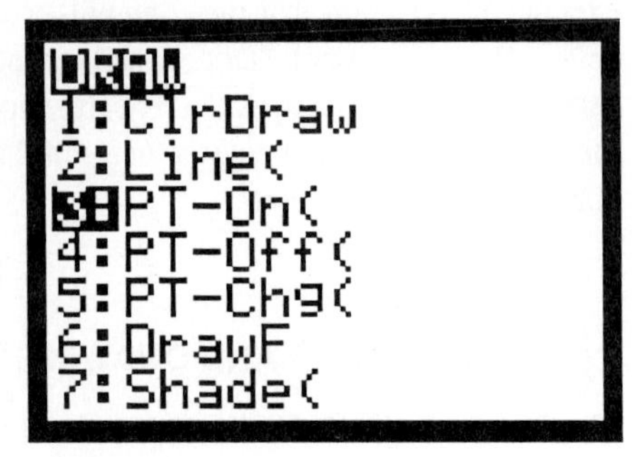

FIGURE 1.42

and any other points you wish to plot. Figure 1.45 shows the **home screen** after all the points from the table have been entered, and Fig. 1.46 shows the accompanying **graphics** screen.

Brenda can now see a trend developing. As the average speed increases, the travel time decreases. The decrease in travel time is very fast between average speeds of 10, 20, and 30 mph, but is much slower when the average speed is in the range of 60 to 90 mph. ◊

EXPLORATION 6: Using the **Line** command from the DRAW menu, draw the triangle with vertices at the points (6, 4), (-5, -1), and (3, -3).

Using the range setting program given in Exploration 4, enter a range factor of 2 to set the viewing rectangle [-9.4, 9.2] by [-6.4, 6.2]. (This viewing rectangle may also be set manually from the RANGE screen.) Select the **Line** command from the DRAW menu to draw a line segment between two specified points in the plane.

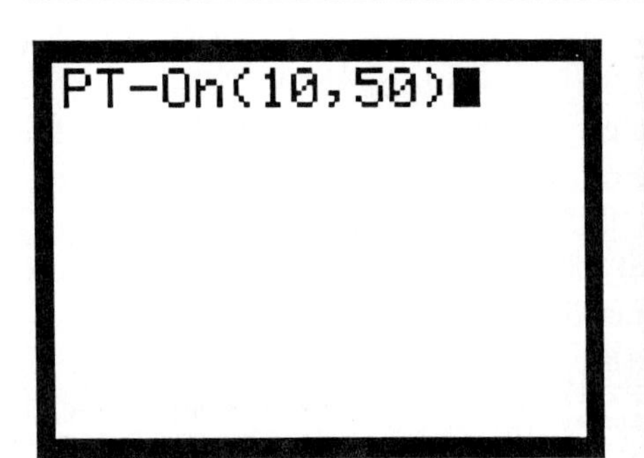

FIGURE 1.43

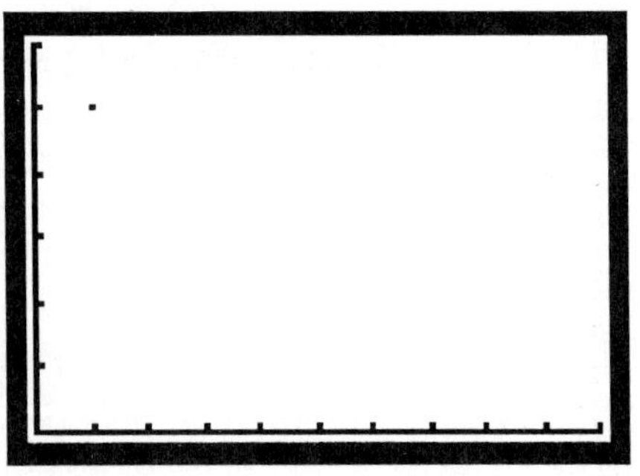

FIGURE 1.44

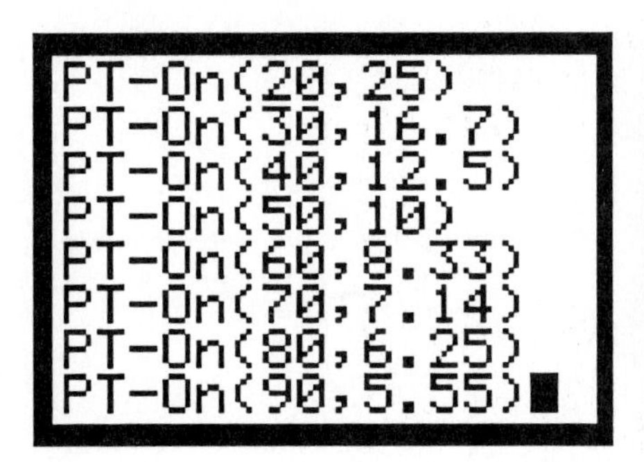

FIGURE 1.45

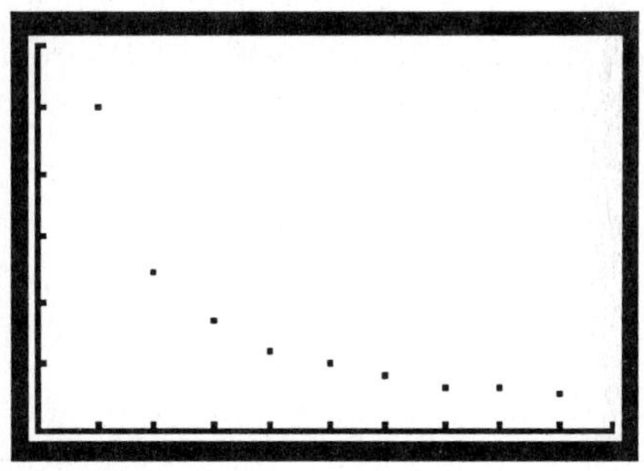

FIGURE 1.46

Enter the x- and y-coordinates of the two points separated by commas in the general form **Line(**x_1, y_1, x_2, y_2**)**, and press [ENTER] to draw the line. Figure 1.47 shows the command for one side of the triangle, and Fig. 1.48 shows the resulting graphics screen.

To continue, press the [CLEAR] key to return to the **home screen**. At this point you may replay and edit the last command to draw the next side of the triangle, or you may select the **Line** command from the [DRAW] menu again. Figure 1.49 shows the commands for all three sides of the triangle, and Fig. 1.50 shows the resulting graphics screen. ◊

EXPLORATION 7: Plot the points from the table in Exploration 5 interactively.

Reset the viewing rectangle to [0, 100] by [0, 60]; clear all equations from the [Y=] menu; activate the **graphics** screen by pressing the [GRAPH] key. Access the [DRAW] menu while the **graphics** screen is visible (active), and select the **3:Pt- On(**

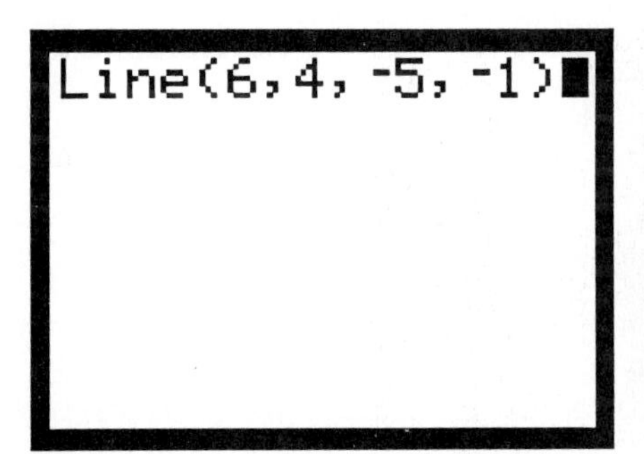

FIGURE 1.47

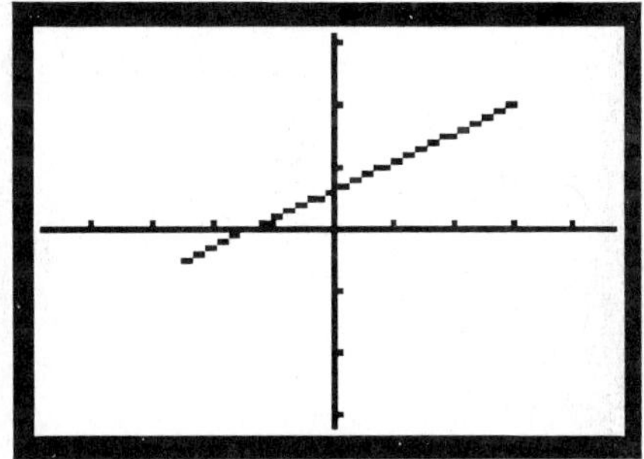

FIGURE 1.48

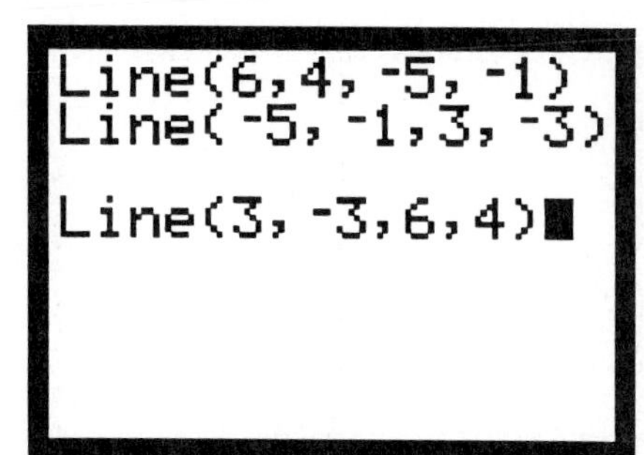

FIGURE 1.49

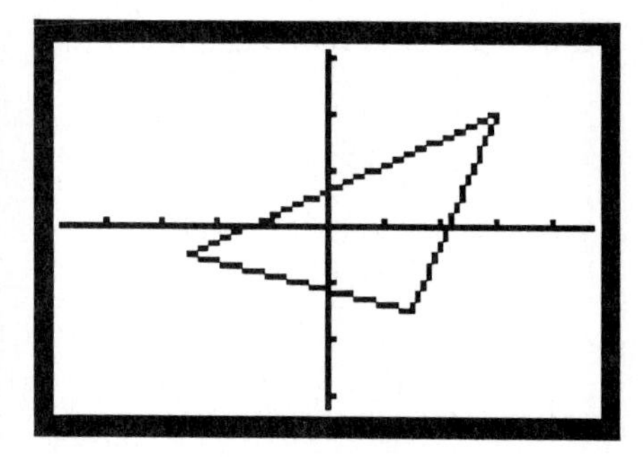

FIGURE 1.50

command. The **graphics** screen will appear with a small flashing cross in the middle of the screen and the coordinates of that point showing at the bottom of the screen. This screen cursor is controlled by the arrow keys. Move the cursor to the first point, (10, 50) (see Fig. 1.51). Notice that you cannot place the cursor exactly on the point (10, 50).

Remember, the screen is a discrete set of 96 horizontal values equally spaced from **Xmin** to **Xmax** and 64 vertical values spaced equally from **Ymin** to **Ymax.** In this case the point (10, 50) is **not** one of the points in the set represented on the screen. The best we can do is select the point as close to (10, 50) as possible. Press ENTER to plot the point. When you move the screen cursor to the next point, (20, 25), you will see the first point plotted on the screen.

When you plot the next point, (20, 25), you will notice that the x-coordinate of 20 is one of the points represented on the screen, but the y-coordinate of 25 must still be an estimate (see Fig. 1.52). Complete the problem by plotting all the other points from the table in Exploration 5. ◊

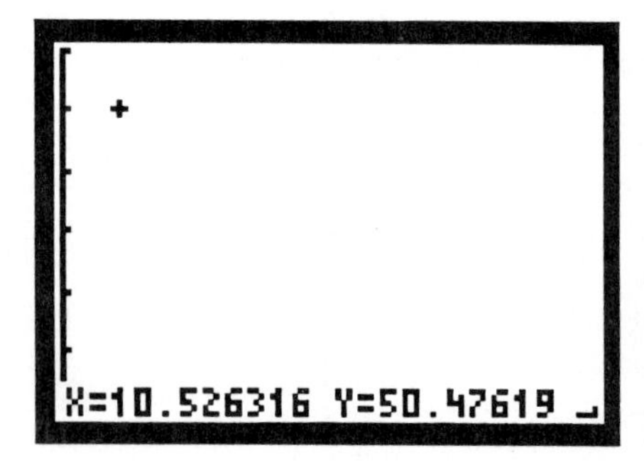

FIGURE 1.51

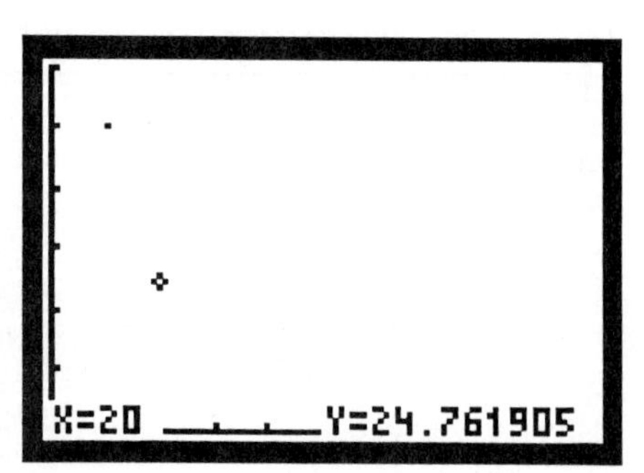

FIGURE 1.52

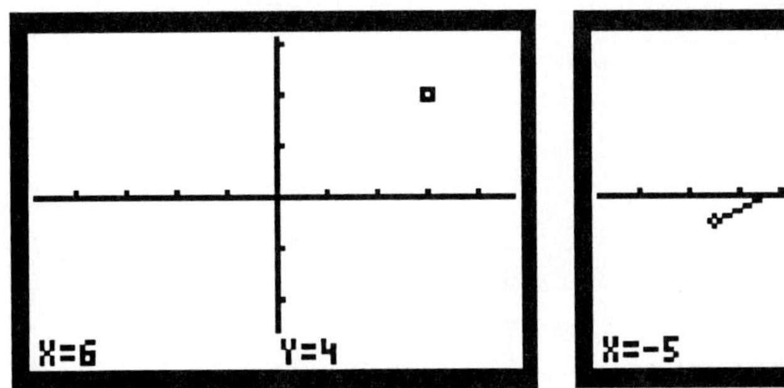

FIGURE 1.53

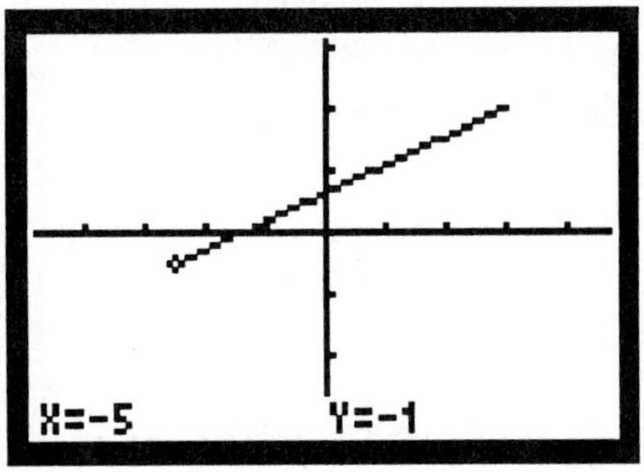

FIGURE 1.54

EXPLORATION 8: Using the **Line** command interactively, draw the triangle given in Exploration 6, with vertices at the points (6, 4), (-5, -1), and (3, -3).

Reset the viewing rectangle to [-9.6, 9.4] by [-6.4, 6.2] using the range program with a factor of 2, or by setting individual values on the RANGE screen. Activate the **graphics** screen and then select the **Line** command from the DRAW menu. A flashing screen cursor will appear at the center of the screen. Move this cursor to one of the vertices of the triangle and press ENTER. The shape of the cursor will change to a small box; you have "pinned down" one end of a line (see Fig. 1.53). Notice that in this viewing rectangle the exact values for the x- and y-coordinates of the vertices are represented on the screen.

Using the arrow keys, "stretch" the line to one of the other vertices and press ENTER to "pin down" the other end of the line (see Fig. 1.54). To draw the next side of the triangle press ENTER to begin the line at the point where you ended the last line and stretch the line to the next vertex. Continue this process to complete the given triangle. ◊

Problems

1. Enter the following pair of linear equations on the Y= menu.

$$Y_1 = -4X + 5 \quad \text{and} \quad Y_2 = .25X - 6.$$

Draw the graph of these two lines in the following viewing rectangles. List an appropriate **Xscl** and **Yscl** for each viewing rectangle.

a. Standard

b. Square

c. [-5, 5] by [-5, 5]

d. first quadrant only

e. first and second quadrant only

f. first and fourth quadrant only

g. [-10, 10] by [-100, 100]

h. [-50, 50] by [-5, 5]

i. [-100, 100] by [-100, 100]

j. [-4.8, 4.7] by [-3.2, 3.1] (friendly range with range factor of 1)

k. [-9.6, 9.4] by [-6.4, 6.2] (friendly range with range factor of 2)

l. [-24, 23.5] by [-16, 15.5] (friendly range with range factor of 5)

m. [-48, 47] by [-32, 31] (friendly range with range factor of 10)

n. first quadrant only, aspect ratio = 1, steps of .1, integer values represented

2. Explain using a diagram why both lines from problem 1 are not visible when the viewing rectangle is [-4.8, 4.7] by [-3.2, 3.1].

3. The Browning family is planning an airplane trip from Minneapolis to Orlando, a distance of 2200 miles.

a. Complete the following table showing how the flight time depends on the average ground speed of the airplane:

Average ground speed	100	200	300	400	500	600	700	800
Flight time								

b. Let the horizontal axis (x) represent the average ground speed and the vertical axis (y) represent flight time. What is an appropriate viewing rectangle that will show the values in the table? What are the appropriate **Xscl** and **Yscl** settings? What is the aspect ratio of your viewing rectangle?

c. Plot the points in your table using the **Pt-On** command numerically from the **home screen.**

d. Expand the table to include average speeds of 150, 250, 350, 450, 550, 650, and 750 mph. Plot these points on the graphics screen interactively.

Average ground speed	150	250	350	450	550	650	750	850
Flight time								

e. Using the **Line** command interactively, connect all of the points on the screen to form a smooth graph.

f. Explain how the flight time is related to average ground time.

g. The pilot announces that the flight time for the trip will be 6.5 hours. Use the graph and the screen cursor to estimate the average speed that the airplane will be traveling.

h. The pilot announces that depending on a tail wind, flight time may be reduced. Using the screen cursor on the graph, estimate the average speed if the flight time was reduced to (i) 5 hours, (ii) 4 hours, (iii) 2 hours.

i. Are there any limits to the average speed this airplane can achieve?

4. Modify the range program listed in Exploration 4 so that it produces friendly viewing rectangles in the first quadrant only.

5. Using the **Line** command interactively, draw a right triangle, an acute triangle, an obtuse triangle, an isosceles triangle, a scalene triangle, an equilateral triangle, and an equiangular triangle. You may wish to clear the graphics screen between each drawing.

6. Using the **Line** command interactively, draw a series of squares embedded within other squares by connecting the midpoints of the sides of a square to form the next smaller square.

7. Using the **Line** command numerically or interactively, draw a regular pentagon, hexagon, and octagon.

8. Using the **Line** command numerically or interactively, draw two rectangles on the graphics screen in such a way that the area of one rectangle is four times that of the other.

9. The Senior class at Shepherd High School is having a candy sale to raise money for a class trip. The wholesale cost of each candy bar is $0.28 and the candy is sold for a cost of $0.50 each. Fixed costs for the project are $500 for postage, telephone, and copying. There are 288 students in the class, and the cost per student for the trip is $50.

a. Make a table showing the number of candy bars sold and the amount of money the class makes.

b. Write an algebraic formula for the amount of money made based on the number of candy bars sold and the fixed cost.

c. Let the horizontal axis represent the number of candy bars sold and the vertical axis represent the total amount of profit the class makes. Set the RANGE screen of the calculator based on your table, and plot the pairs of values from the table using the **Pt-On** command.

d. Use the **Line** command interactively to connect the points on the graphics screen. Is there a trend visible on the graphics screen?

e. If candy bars are packed 48 to a case, how many cases must be sold to break even (make up for fixed costs)? How many cases must be sold to earn enough money for the class trip?

10. Set the graphics screen to the **Standard** range. Using the **Line** command interactively, draw the line with the equation $y = -2x - 4.5$. Graph this line using the **DrawF** command from the DRAW menu to confirm your estimate. (The command should read **DrawF -2X – 4.5**; press ENTER to draw the graph.)

FOR ADVANCED ALGEBRA STUDENTS

11. Draw the graph of the line $y = x + 1$ in the **Standard** viewing rectangle.

- **a.** What is the slope of this line? What angle should this line make with the x axis? Does the **Standard** viewing rectangle give the correct angle?
- **b.** Draw the graph in the viewing rectangle [-10, 10] by [-100, 100]. Explain why the graph looks so flat. Is the slope of this line 1? Explain.
- **c.** Draw the graph again by selecting the **Square** option from the [ZOOM] menu. What is the size of this viewing rectangle?
- **d.** Draw the graph in the viewing rectangle [-100, 100] by [-10, 10]. Is the slope of this line 1? Explain.
- **e.** Draw the graph again by selecting the **Square** option from the [ZOOM] menu. What is the size of this viewing rectangle? Is the same viewing rectangle found in part (c)? Explain.

12. Plot the following set of points using the **Pt-On** command:

x	-3	-2	-1	0	1	2	3	4	5	6	7
y	36	25	16	9	4	1	0	1	4	9	16

- **a.** Connect the points on the screen using the **Line** command interactively.
- **b.** Find the algebraic formula that represents this set of points. Use the **DrawF** command from the [DRAW] menu to check your formula.

13. The graph of the equation $x^2 + y^2 = 16$ is a circle centered at the origin with a radius of 4 units. This equation does not represent a function since the graph fails the vertical line test. To graph this relationship on the TI-81, solve the equation for y in terms of x and enter the resulting two equations (i.e. + and -) separately in Y_1 and Y_2.

- **a.** Draw the graph of the circle in the **Standard** viewing rectangle. Describe the graph. Are there any problems?
- **b.** Select the **Simul** mode from the [MODE] screen and draw the graph again. What happened?
- **c.** Select the **Square** option from the [ZOOM] menu to draw the graph of the circle again. Describe the graph. Are there any problems?
- **d.** Use the range program to set a friendly viewing rectangle with factor 1 (i.e. [-4.8, 4.7] by [-3.2, 3.1]), and draw the graph of the circle. Describe the graph. Are there any problems?
- **e.** Draw the graph of the circle using the range program with a factor of 2 (i.e. [-9.6, 9.4] by [-6.4, 6.2]). Describe the graph. Are there any problems?

14. Draw the graph of a circle centered at the origin and a circle centered at the point (1, -2), both with radius 4, in a viewing rectangle that shows a visually correct graph of both complete circles.

15. The point (1, 1) is on the circle $(x+2)^2+(y+3)2=25$. Draw the graph of this circle in an appropriate viewing rectangle, and plot the center of the circle using the **Pt-On** command. Calculate the equation of a line that will represent the diameter of this circle. Draw the graph of the circle and the line.

16. The point (-5, 1) is on the circle $(x+2)^2+(y+3)2=25$. Draw the graph of this circle and plot the center. Calculate the point on the circle directly opposite the point (-5, 1). Draw the diameter of the circle from (-5, 1) to the opposite point using the **Line** command.

17. Modify the range program given in Exploration 4 so that you may enter the coordinates of the center of the screen as well as the range factor, and the resulting viewing rectangle has all of the properties of those produced by the original program. (These properties include an aspect ratio of 1, steps of $\frac{1}{10}$ of the range factor, and that values in the set include the integers within the viewing rectangle.) *Hint:* Think of this process as a horizontal and vertical translation of every point in the viewing rectangle.

18. Using the range program, set a friendly viewing rectangle centered at the origin with a range factor of 2 (i.e. [-9.6, 9.4] by [-6.4, 6.2]). Draw the triangle with vertices (-7, -1), (-1, -4), and (3, 4) using the **Line** command.

    a. Prove that the given triangle is or is not a right triangle by using the distance formula.

    b. Prove that the given triangle is or is not a right triangle by a slope argument.

    c. Find the equations of the lines represented by the three sides of the triangle, and draw the triangle by drawing the graphs of these three lines.

    d. Find the midpoint of each side of the triangle.

    e. Connect the midpoint of each side to the opposite vertex of the triangle (forming the **medians** of the triangle) using the **Line** command.

    f. Prove that all three medians of the triangle intersect at the same point.

    g. **Extra Challenge:** Write a program that will draw a triangle given the coordinates of the three vertices and that will prove whether the figure is or is not a right triangle by a slope argument and a distance argument.

19. Using the **Line** command, draw the quadrilateral with vertices at (-4, -2), (-6, 4), (7, 5), and (5, -5).

    a. Prove that this figure **is not** a parallelogram.

    b. Locate the midpoints of each of the four sides and connect these points with line segments. What figure is formed?

    c. Prove that the figure formed by connecting the four midpoints **is** a parallelogram by a slope argument and a distance argument.

20. Plot the points (-8, -6), (2, 2), and (6, 5) and connect them with line segments. Do these three points appear to lie on the same line? Test these points algebraically to see if they are collinear.

1.4 Exploring Lines and Linear Functions

GRAPHING LINEAR FUNCTIONS

A ***function*** is a mathematical relationship consisting of pairs of points (x, y) in which no two points have the same first coordinate. For example, the pairs of points (-2, 4), (-1, 1), (0, 0), (1, 1), and (2, 4) would constitute a function. Notice that some (or all) pairs of points may have the same second coordinates, but never the same first coordinate. A ***linear function*** is a functional relationship whose graph is a straight line. All pairs of points represented by the function definition lie on a straight line.

EXPLORATION 1: Use the **Pt-On** command to plot the set of points in Table 1.1. Clear all functions from the [Y =] menu and set an appropriate viewing rectangle. Figure 1.55 shows the graphics screen with the points plotted. The viewing rectangle is [-9.6, 9.4] by [-6.4, 6.2].

Notice that the points in this set form a straight line. If we were to add more points from this functional relationship, they would also fall on the same straight line. Add the points (-2.4, 3.8), (-1.2, 1.4), (-0.6, 0.2), (0.6, -2.2), (1.4, -3.8), and (2.2, -5.4) to the graphics screen. Each new point falls on the same line. The linear function that represents **all** the points on the line is $\mathbf{y = -2x - 1}$. For any value of x you can calculate a corresponding value for y that completes a pair of coordinates (x, y). When all of these coordinate pairs are plotted on the graph, a straight line is graphed.

Use the **DrawF** command (for "draw the function") from the [DRAW] menu to draw the linear function $y = -2x - 1$. Figure 1.56 shows the command entered on

TABLE 1.1

x	y
-3	5
-2	3
-1	1
0	-1
1	-3
2	-5

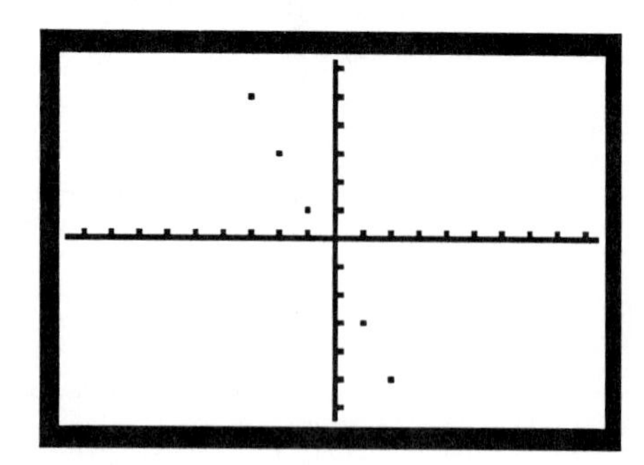

FIGURE 1.55

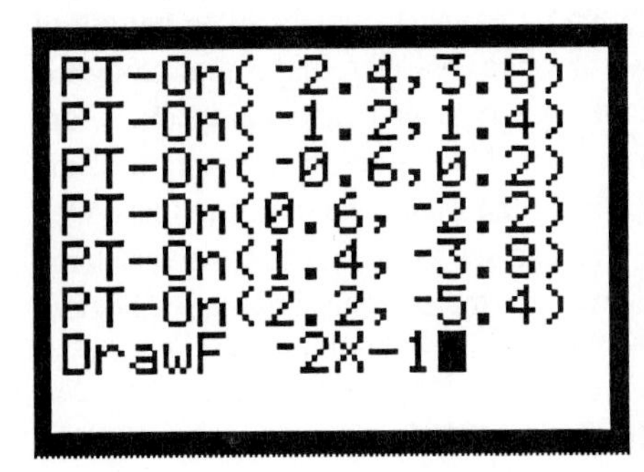

FIGURE 1.56

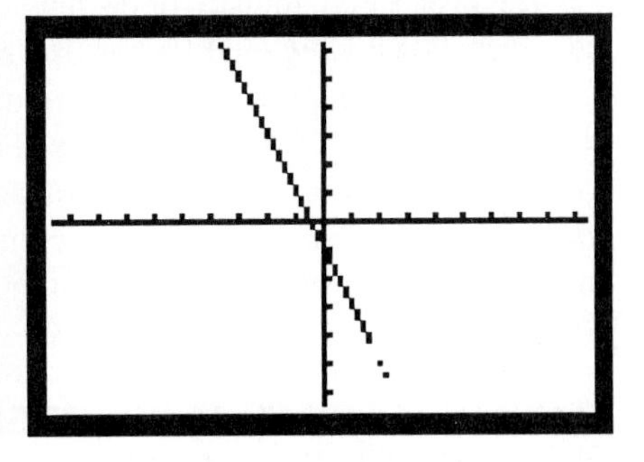

FIGURE 1.57

the **home screen** and Fig. 1.57 shows the resulting graphics screen. (Carefully note the difference between the opposite of a number symbol [(–)] and the subtraction symbol [—]). Do not enter "Y=" before the function definition. To execute the **DrawF** command, press the [ENTER] key. Watch closely as the line connects all the points you have plotted on the screen into one line.

We can also draw the graph of this linear function using the [Y=] menu. Activate this menu by pressing the [Y=] key. Clear any other functions from this menu and enter $\mathbf{Y_1 = -2X - 1}$. With your viewing rectangle set at [-9.6, 9.4] by [-6.4, 6.2], press the [GRAPH] key to draw the graph of this linear function. When the [Y=] menu is changed, the graphics screen was automatically cleared and the new graph was drawn as before (see Fig. 1.57). When graphing functions from the [Y=] menu, the TI-81 remembers the function definition until you erase or change it. Press the [TRACE] key. A small flashing box will appear on the graph near the horizontal center of the screen (see Fig. 1.58).

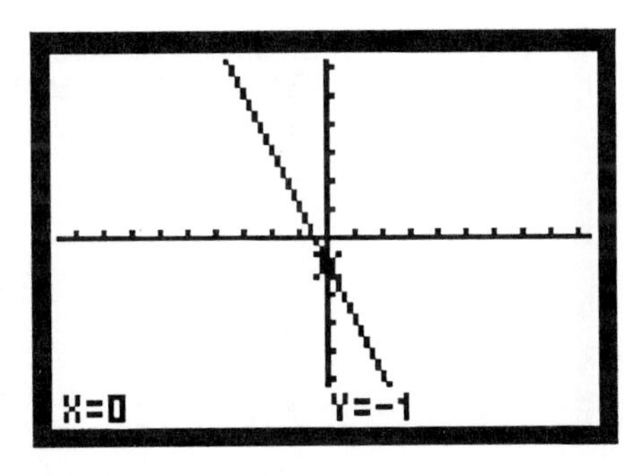

FIGURE 1.58
[TRACE] cursor

The x- and y-coordinates of the highlighted point are displayed at the bottom of the screen. Use the ◀ and ▶ arrow keys to move the TRACE cursor along the function. Notice that the points you plotted on the screen, and many more, are represented by the TRACE cursor (see Fig. 1.59). The values at the bottom of the screen represent rows from a table of values giving the x and y values of points on the graph. If you continue to scroll right or left, you can see these values for a very large number of points, even if the TRACE cursor leaves the current viewing rectangle. ◊

STANDARD FORM VS. SLOPE-INTERCEPT FORM OF THE EQUATION OF A LINE

The **standard form** of the equation of a linear function is $ax + by = c$, where a, b, and c are integers. The TI-81 is a function grapher that can only deal with equations of the form y = "some function of x." This means that you must solve a linear equation given in standard form for y in terms of x. For example, given the standard form equation $3x + 4y = 12$, by solving for y we get

$$\begin{aligned} 3x + 4y &= 12 \\ 4y &= -3x + 12 \\ y &= -.75x + 3 \end{aligned}$$

We now have the equation in the form $y = mx + b$, which can be entered in the TI-81. This is called the **slope-intercept form** of the equation of a line.

SLOPE AND Y-INTERCEPT

The slope of a line is a measure of the **inclination** or **steepness** of a line. It is an important concept in the study of the behavior of functions of all types. The slope, m,

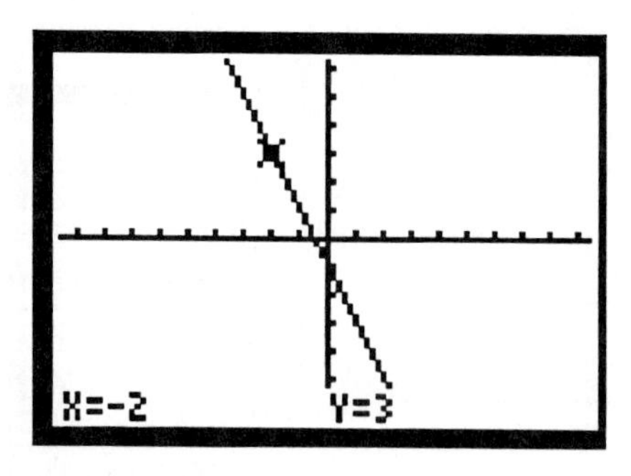

FIGURE 1.59
Move along the graph

of a line is computed using the formula $m = \frac{\Delta y}{\Delta x}$, where Δy is the change in y and Δx is the corresponding change in x. If an equation of a line is in the form $y = mx + b$, then the coefficient m is the **slope** of the line and b is the ***y*-intercept** (see Fig. 1.60). The y-intercept of a line is the point at which the line crosses, or "intercepts," the y-axis. Since the y-axis lies at the point $x = 0$, the first coordinate of the y-intercept is always zero.

EXPLORATION 2: Graph $y = x$, $y = 2x$, and $y = 4x$ in the standard viewing rectangle.

Enter the three equations on the menu as $\mathbf{Y_1}$, $\mathbf{Y_2}$, and $\mathbf{Y_3}$. Select **Standard** from the [ZOOM] menu. Figure 1.61 shows the resulting graphs. It is apparent from the graph that the line $y = 4x$ is steeper than the line $y = 2x$. In general, if $m > n > 0$, then $y = mx$ is steeper than $y = nx$.

The line $y = x$ should bisect the right angle formed by the coordinate axes. A careful inspection of Fig. 1.61 does not support this mathematical fact. The problem is that the aspect ratio of the graphics screen is not 1. To overcome this problem, select **Square** from the [ZOOM] menu or set a "friendly" viewing rectangle manually or by using a program. ◊

EXPLORATION 3: Continue to explore the idea of the slope of a line by adding more graphs of linear functions of the form $y = mx$ using various fractional and negative values for m.

Add a fourth linear function to the [Y=] menu: $\mathbf{Y_4 = 10X}$. Predict where this line will graph before you press [GRAPH]. Were you correct? With a coefficient of 10, this line will be steeper than the line $y = 4x$. Also notice that the graphics screen is cleared when any change is made on the [Y=] menu.

To add more graphs to the graphics screen without clearing the present graphs, use the **DrawF** command. The graphs already on the screen will not be cleared; the

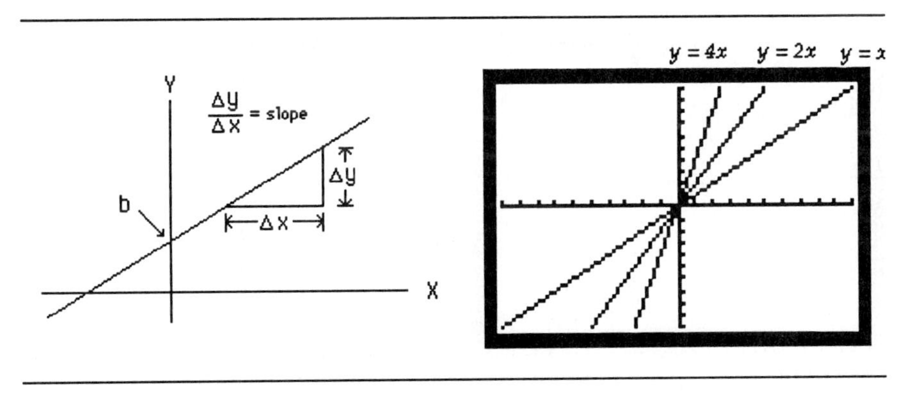

FIGURE 1.60

FIGURE 1.61

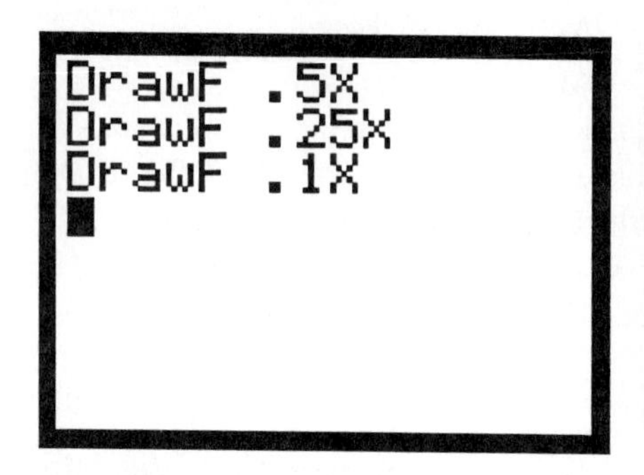

FIGURE 1.62

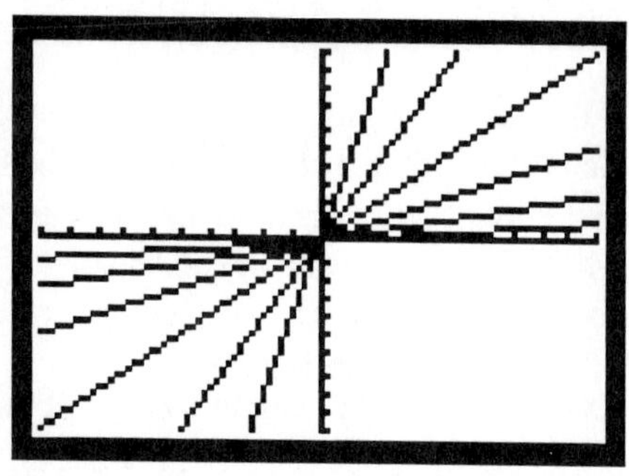

FIGURE 1.63

new graphs will be drawn on the same screen. Add the graphs of $y = .5x$, $y = .25x$, and $y = .1x$ to the graphics screen one at a time. Figure 1.62 shows the **home screen** after the three commands have been executed, and Fig. 1.63 shows the resulting graphics screen with the graphs of seven linear functions.

What happens if the value of m is negative? Add the graphs of $y = -x$, $y = -2x$, $y = -4x$, and $y = -10x$. What about fractional values that are negative, like -.5, -.25, or -.1? Make a general rule that you can use to predict how a linear equation will look based on the value of m for the slope of the line. ◊

EXPLORATION 4: Explore the effect of changing the value of the y-intercept, b, in the slope-intercept form of a line.

Figure 1.64 shows four equations entered on the [Y=] menu, and Fig. 1.65 shows the resulting graphics screen for the [-9.6, 9.4] by [-6.4, 6.2] viewing rectangle.

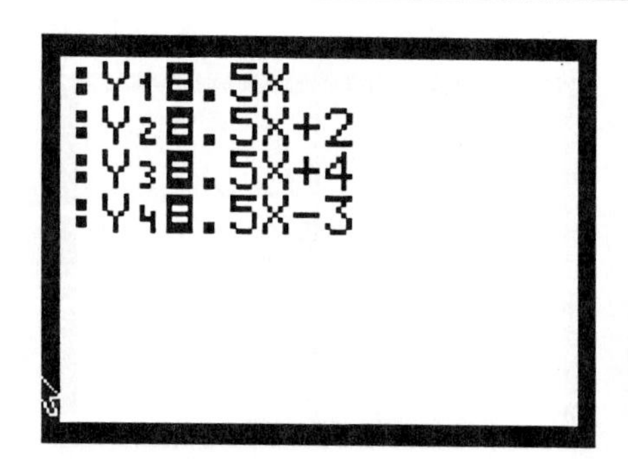

FIGURE 1.64

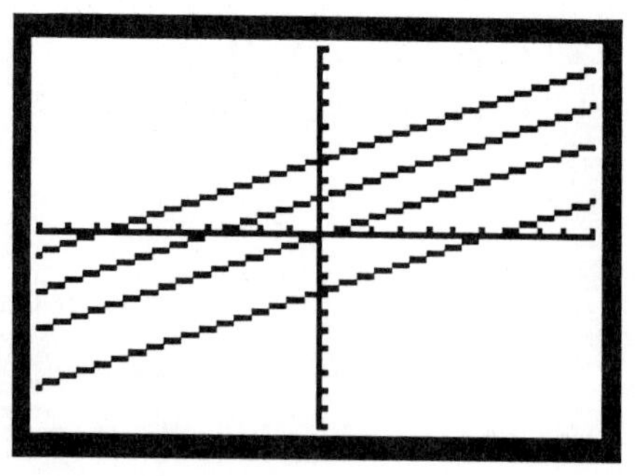

FIGURE 1.65

Can you make a general rule to predict the position of any equation of the form $y = .5x + b$? ◊

APPLICATION EXPLORATION 1: The model $C = \frac{5}{9}(F - 32)$ is the Fahrenheit (F) to Celsius (C) temperature conversion formula. Draw a graph of this model. Is the graph a straight line? Why? What is its slope? What does the slope mean in this problem situation? Use a graph to estimate the temperature Celsius if the present outside temperature is 98°F.

The graph of $C = \frac{5}{9}(F - 32)$ can be plotted on the TI-81 by graphing $\mathbf{Y_1} = \frac{5}{9}(\mathbf{X} - 32)$. In this form of the equation, $C = \mathbf{Y_1}$ and $F = \mathbf{X}$. You may use this form of the equation or expand the equation into slope-intercept form using the distributive property:

$$\mathbf{Y_1} = \frac{5}{9}(\mathbf{X} - 32)$$

$$\mathbf{Y_1} = \frac{5}{9}\mathbf{X} - \frac{5}{9}(32)$$

$$\mathbf{Y_1} = \frac{5}{9}\mathbf{X} - \frac{160}{9}$$

Figure 1.66 shows the two forms of the equation entered on the [Y=] menu; only one of these is necessary to draw the correct graph. Figure 1.67 shows the resulting graph in the viewing rectangle [-100, 100] by [-100, 100] with **Xsxl** = **Yscl** = 10. The x-axis represents the temperature in degrees Fahrenheit and the y-axis represents the temperature in degrees Celsius.

The graph is a straight line with slope $\frac{5}{9}$ and a y-intercept of $\frac{-160}{9}$. In this model,

FIGURE 1.66

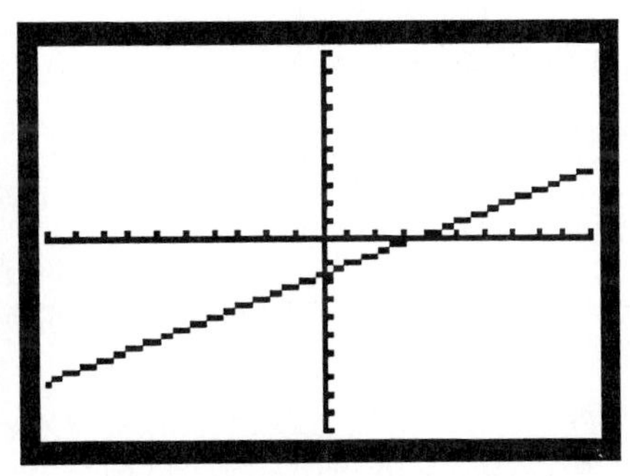

FIGURE 1.67

a slope of $\frac{5}{9}$ means that for every 9° change in the Fahrenheit temperature, there is a corresponding change of 5° in the Celsius temperature.

To use the graph to estimate Celsius temperature given the Fahrenheit temperature, first press the [TRACE] key, and use the [◀] and [▶] arrow keys to move the trace point to where the display reads $x = 98.421053$ and $y = 36.900585$. This means that a temperature of approximately 98.4°F is equivalent to approximately 36.9°C. Now press the keys [ZOOM] **[8: Integer]** [ENTER] [TRACE]. Notice you have adjusted the viewing rectangle so the trace point for the x values (Fahrenheit temperature) read out as integers (i.e. 95, 96, 97, etc.). Move the trace point using the [◀] and [▶] arrow keys to confirm this. Now you can see that 98°F is equivalent to 36.666667°C. This should lead you to suspect the exact equivalent Celsius measure is $36\frac{2}{3}$°. Of course, you may confirm this by computing $C = \frac{5}{9}(98 - 32) = \frac{330}{9} = \frac{110}{3} = 36\frac{2}{3}$. ◊

APPLICATION EXPLORATION 2: Central Lumber Co. charges a fee of $5.00 plus 10% of the purchase price to deliver material to your home. Make a graph showing the total cost of delivery as a function of the amount of material purchased. Let the horizontal axis represent the amount of the purchase and the vertical axis represent the delivery cost.

(a) Using your graph, find the delivery charge for a purchase of $450.

Since the delivery cost is $5 plus 10% of the amount purchased, we can write the linear equation $y = .10x + 5$, where y represents the cost of delivery and x represents the amount purchased. Enter this equation on the [Y=] menu. Set the range based on the parameters of the problem. The amount purchased varies from $0 to about $500. The delivery fee could be estimated at between $0.00 and $55.00 (10% of $500 + $5). Figure 1.68 shows the graph of the function $\mathbf{Y_1 = .10X + 5}$ in the viewing rectangle [0, 475] by [0, 55]. **Xmax** is set at 475 because this value gives horizontal steps of 5 units for each pixel, since 475/95 = 5.

Use the [TRACE] function to find $450 as the purchase price. The y values give the delivery fee (Fig. 1.68). For this amount of material, the delivery cost is $50.

(b) Using your graph, find the cost of materials if the delivery cost was $40.50.

Move the [TRACE] cursor until the y-coordinate of the point reads $40.50; the x-coordinate is 355, representing material costs of $355 (see Fig. 1.69). ◊

APPLICATION EXPLORATION 3: Corey Winston has $16,000 to invest. He decides to invest the money in two parts, one paying 8.4% and the other paying 11.2% per year.

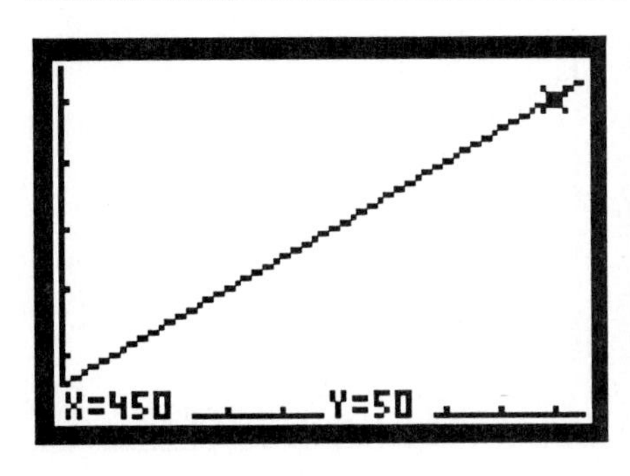

FIGURE 1.68

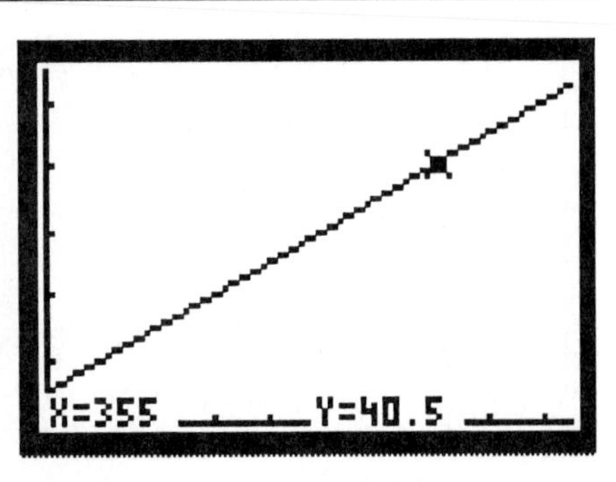

FIGURE 1.69

(a) Write an algebraic representation of the problem situation. Let x represent the amount of money invested at 8.4% interest, so the yearly interest earned on this money is given by $.084x$. If x is the amount of money invested at 8.4%, then $(16{,}000 - x)$ is the amount invested at 11.2%. The yearly interest on this part of the money is given by $.112\,(16{,}000 - x)$. The total interest earned, y, is the sum of the two parts, or $y = .084x + .112\,(16{,}000 - x)$.

(b) Draw a graph showing how the total amount of interest earned in one year depends on the amount of money deposited at 8.4%. Enter the algebraic representation on the Y= menu in the form just given, or in slope-intercept form found by expanding and collecting like terms:

$$y = .084x + .112\,(16{,}000 - x)$$
$$y = .084x + 1792 - .112x$$
$$y = -.028x + 1792$$

Set the RANGE based on the values in the problem. Since x represents the amount of money invested at 8.4%, this may range from \$0.00 to \$16,000. Set **Xmin** = 0 and **Xmax** = 16,000. Choose an appropriate **Xscl** such as 1000. The y values represent the total interest earned in one year. If \$0.00 were invested at 8.4% interest, the total interest earned would be \$1792. Why? If all \$16,000 were invested at 8.4%, the interest would be -.028(16,000) + 1792 = \$1344. Set **Ymin** = 1200 and **Ymax** = 1800. Again, choose an appropriate **Yscl** of 50 or 100. Figure 1.70 shows the screen with the graph drawn and the TRACE cursor activated. Why is the slope negative?

(c) If the total yearly interest is \$1612.80, how much money was deposited at the 8.4% rate? The y values represent the total yearly interest. Move the TRACE cursor until the y values on the screen approach \$1612.80. The x value corresponding to this amount of total interest is \$6400, meaning that if \$6400 was

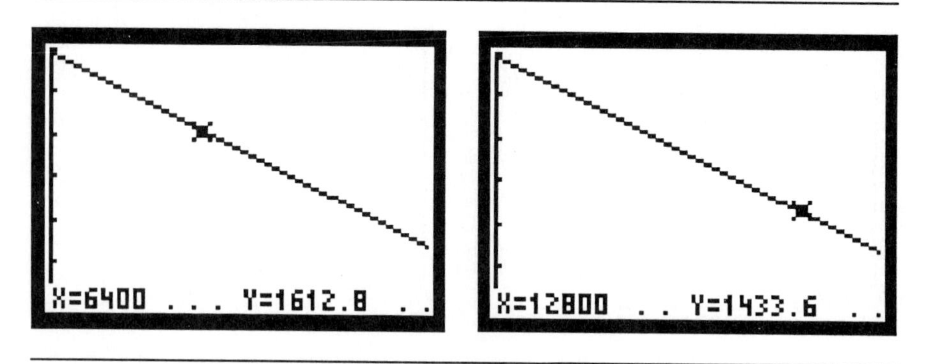

FIGURE 1.70

FIGURE 1.71

invested at 8.4% and \$9600 (\$16,000 – 6400) was invested at 11.2%, then the total interest earned in one year by both parts is \$1612.80 (see Fig. 1.70).

(d) If \$3200 is deposited at 11.2% interest, what is the total amount of interest for one year? If \$3200 is invested at 11.2%, then \$16,000 – 3200 = \$12,800 was invested at 8.4% interest. Move the TRACE cursor until the x value approaches 12,800. The y value at this point is 1433.6, meaning that the total yearly interest is \$1433.60 when \$3200 is invested at 11.2% interest (see Fig. 1.71).

(e) What is the most interest that can be earned in one year? What is the least? Explain how these two amounts are found. This part of the problem has already been solved! When we considered possible values for the total amount of interest earned to set **Ymin** and **Ymax**, we calculated the highest and lowest amount of interest which could be earned. Intuitively, the most interest will be earned if we place all of the \$16,000 in the account paying 11.2% interest and no money in the account paying the lower 8.4% interest. This corresponds to the equation $y = -.028(0) + 1792 = 1792$. Likewise, the lowest amount of interest is earned when all of the \$16,000 is placed in the account paying 8.4% interest. This corresponds to the equation $y = -.028(16000) + 1792 = -448 + 1792 = 1344$. These are the same values we used to set the scale on the vertical (y) axis. ◊

Problems

1. Use the **Pt-On** command to plot the following sets of points. Then use the **Line** command interactively to see if they form a straight line. Set the viewing rectangle to [-9.6, 9.4] by [-6.4, 6.2].

a.

x	y
-3	-1
-2	0
-1	1
0	2
1	3
2	4

b.

x	y
-2	-5
-1	-3
0	-1
1	1
2	3
3	5

c.

x	y
-2	3
-1	2
0	1
1	0
2	-1
3	-2

d.

x	y
-2	-2
-1	-1
0	1
1	2
2	3
3	5

2. If the sets of points in the previous problem do form a straight line, then there is a functional relationship. Write the function, and test it using the **DrawF** command.

3. Use **DrawF** to draw the following functions and from the interactive mode find at least two other points on the line. Test the points by substituting them into the function. Did the points fit the equation exactly? Why or why not?

   a. $y = 4x + 3$ **b.** $y = -3x + 2$

4. Graph the functions from the previous problem in the [Y=] menu, and use the [TRACE] cursor to find at least two other points on the lines. Test the points by substituting them into the functions. Did the points fit the equation exactly? Why or why not?

5. Select **Grid On** command from the [MODE] menu. Now when you do any graphing, a grid of points will appear on the graphics screen. Use the **Line** command interactively to draw the lines connecting the following pairs of points, and calculate the slope $\frac{\Delta y}{\Delta x}$ as you would on graph paper.

   a. (1, 2) and (3, 4) **b.** (-1, 2) and (2, 3)

   c. (1, -3) and (4, -4) **d.** (0, 3) and (4, 0)

6. Turn on the **Grid On** option from the [MODE] menu to simulate graph paper. In a friendly viewing window, use the grid and the interactive mode to draw a line with the following slopes through the point (1, 1):

   a. $m = 2$ **b.** $m = -3$

   c. $m = \frac{2}{3}$ **d.** $m = -\frac{3}{4}$

7. Enter the following functions on the [Y=] menu and predict where the lines will graph and whether they will increase or decrease as you move from the left to the right. What effect does the number before the X have?

   a. $y = 4x$ **b.** $y = \frac{2}{3}x$ **c.** $y = -x$

   d. $y = -.5x$ **e.** $y = 100x$ **f.** $y = -2.5x$

8. Enter the following functions into the [Y=] menu and predict the effect of the constant value on the function. Using a friendly viewing rectangle will give you an aspect ratio of 1.

a. $y = x + 1$ **b.** $y = x - 1$

c. $y = x + 3$ **d.** $y = x - 5$

9. From your knowledge of slope and y-intercepts, graph the following lines without using the TI-81. Confirm your paper and pencil graphs by drawing the same graphs on the TI-81 in a friendly viewing window. Use the [TRACE] cursor to locate the y-intercepts of the following functions and make a conclusion about the y-intercept of any linear equation.

a. $y = 2x + 2$ **b.** $y = -3x - 5$

c. $y = .5x + 1.5$ **d.** $y = 10x - 1$

10. In a friendly viewing window, graph the following pairs of lines to investigate what happens when the lines have the same slope. Make a conjecture about the graphs of lines with the same slope and test this conjecture with other pairs of lines.

a. $y = 3x + 3$ and $y = 3x - 3$

b. $y = -2x + 1.5$ and $y = -2x - 8.1$

c. $y = 1.3x - 2$ and $y = 1.3x + 6$

11. In a friendly viewing window, graph the following pairs of lines to investigate what happens when the lines have slopes that are negative reciprocals. Make a conjecture about the graphs of lines with negative reciprocal slopes and test this conjecture with other pairs of lines.

a. $y = \frac{1}{2}x + 5$ and $y = -2x - 2$

b. $y = -\frac{4}{5}x - 1$ and $y = \frac{5}{4}x + 1$

c. $y = x - 3$ and $y = -x - 2$

12. Remember that you are using a function grapher. Rewrite the following so that they could be entered on the [Y=] menu or graphed using the **DrawF** command. Are they increasing or decreasing functions as you move from left to right? From the general form $Ax + By = C$, make a conjecture about the slopes of the lines and confirm this algebraically.

a. $x + y = 1$ **b.** $x - y = 3$

c. $2x + 3y = 6$ **d.** $3x - 4y - 5 = 0$

e. $x - 2y = 1$ **f.** $3y - 4x = 12$

13. In the friendly viewing window use the **Line** command interactively to investigate the change of the x- and y-intercepts by graphing lines with different negative slopes through the point (1, 1). What happens as the slope increases? What happens to the x-intercept as the y-intercept decreases? What happens to the area of the right triangle that is formed? What slope would maximize this area?

14. Given are the circumferences (to the nearest hundredth) and the diameters of four circles. Plot them using **Pt-On** so the diameter is on the x-axis and the circumference is on the y-axis. Use the **Line** command algebraically to connect the points after setting an appropriate range. Are they collinear? What is the slope of the line? What function would be used to connect these points? Explain.

a. $d = 3$ and $C = 9.42$ **b.** $d = 5$ and $C = 15.71$

c. $d = 12$ and $C = 37.7$ d. $d = 15$ and $C = 47.12$

15. The formula for the cost of producing x items is $C = px$, where p is the price per item. Assuming a cost of $3.25 per item we have a functional relationship between the number of items and the total cost.
    a. Write a function with y = cost and x = the number of items.
    b. Graph the function using the [Y=] menu. Which [RANGE] values make sense in this problem situation?
    c. Why does the graph pass through the origin?
    d. What meaning does the slope have in this problem?
    e. Use [TRACE] to find the cost of 9 items.
    f. Use [TRACE] to find the number of items that can be purchased for $150.
16. Suppose in the previous problem, there was an overhead of $500. How would this alter your answers to parts (a)–(f) in problem 15?
17. The formula for distance traveled is $\boldsymbol{d = rt}$, where d is the distance, r is the rate of speed, and t is the time. For the following assume that you are traveling at an average rate of 30 mph.
    a. Write a function to show the relationship between the distance traveled (y) and the time it takes (x).
    b. Graph the function in (a). What [RANGE] values make sense in the problem?
    c. Why does the graph intercept at the origin?
    d. What does the slope mean in this problem?
    e. Use [TRACE] to find the distance you would travel in 1.8 hours.
    f. Use [TRACE] to find the time it would take to travel 87 miles.
    g. Suppose that you had already traveled 10 miles. How would this change the previous questions?
18. The cost for a long distance phone call is $1.10 for the first minute and $0.25 for each additional minute.
    a. Write a function y to represent the cost of a call taking x minutes.
    b. Graph the function using the [Y=] menu. What [RANGE] values make sense in this problem situation?
    c. What meaning does the y-intercept have in this problem?
    d. What meaning does the slope have in this problem?
    e. Use [TRACE] to find out how much a 5-minute call costs.
    f. Use [TRACE] to find out how long a person could talk for $7.50.
19. The formula for the temperature (T) at different altitudes is $T = t - d/150$, where t is the ground temperature in degrees Celsius and d is the altitude in meters. Assume a ground temperature of 26 degrees Celsius so we can graph the relationship between different altitudes (d) and different temperatures (T).
    a. Write a function $\mathbf{Y = mX + b}$, where $\mathbf{Y}$ is the temperature T and $\mathbf{X}$ is the altitude d.
    b. Choose a proper [RANGE] and graph the function.
    c. What does the Y-intercept mean in this problem?

d. What does the slope mean in this problem?

e. Use TRACE to find the temperature at an altitude of 900 meters.

f. Use TRACE to find the altitude if the temperature is 15 degrees Celsius.

20. The TRACE cursor can be used as a "counter" as well as a tool for finding specific values on a line. Set your RANGE to [0, 95] by [0, 63], so that when you TRACE you will only get whole numbers. Then graph **Y = 9X** and TRACE to see that any whole number divisible by 9, has 9 as the sum of its digits if we repeatedly add the digits. You can TRACE until you are convinced. Inductively show the divisibility test for 6 in the same way.

FOR ADVANCED ALGEBRA STUDENTS

21. We know that the graphing calculator in its present mode can only graph functions. Can you think of a way to graph a vertical line like **X = 3**?

22. Use linear functions to draw a 45° right isosceles triangle and a 30°-60° right triangle.

23. Use the **Line** command interactively to draw a line from one point to the other. Now try to locate the midpoint, and check your answer algebraically.

24. Use the **DrawF** command from the DRAW menu or the Y= menu to graph the following functions. What type of quadrilateral is formed? Use a friendly viewing window to help in the visualization and check algebraically.

a. $4x + 3y = 14$ **b.** $4y - 3x = -23$

c. $3y + 4x = 39$ **d.** $4y - 3x = 2$

25. Using the following points, find the functions that would draw the quadrilateral ABCD and its two diagonals, and graph them. Do the diagonals appear to be perpendicular? If so, what kind of quadrilateral is it? Check algebraically.

A(1, 2) B(5, 7) C(9, 2) D(5, -3).

26. In a friendly viewing rectangle with an aspect ratio of 1, write and graph the equation of the line that passes through the points (0, 0) and $(1, \sqrt{2})$. Turn **Grid On** from the MODE menu and use TRACE to convince yourself that there are no other integer values on the line except (0, 0). You may want to use **Zoom In** (ZOOM menu) to convince yourself. Keep changing the RANGE to larger values, say 10–20 and then 20–30, etc. Explain why this is true for any radical that is not the radical of a perfect square.

27. On the Y= menu use 2nd ABS keys to graph the functions $y = |x| - 6$ and $y = -|x| + 6$ in a friendly viewing window. What type of quadrilateral is formed? Prove it algebraically.

28. For the following division problems, graph them as one function making sure there are parentheses in both the numerator and the denominator. Use the y-intercept and the slope of the line to write the answer, and check algebraically using long division. Explain the "hole" in the graph in parts (a) and (c), and why there are no "holes" in the graph in parts (b) and (d).

a. $y = \dfrac{x^2 + 8x + 12}{x + 2}$ **b.** $y = \dfrac{6x^2 - 13x + 6}{3x - 2}$

c. $y = \dfrac{3x^3 - 19x^2 + -12x + 76}{x^2 - 4}$

d. $y = \dfrac{4x^3 + 12x^2 + 7x + 5}{2x^2 + x + 1}$

EXPLORING STATISTICS

29. Statistically, we know that a linear function may not adequately describe our data. Even though the TI-81 has a statistics menu (see the *Exploring Statistics with the TI-81* in this series), we are going to draw the line graph using the **Line** command either algebraically or interactively. Labeling the x-axis as the *day* and the y-axis the *number of students present*, make a line graph of the number of students present in your math class this week.

30. There are easier ways to graph a scatter plot, but for now we will use the **Pt-On** command in the DRAW menu. Using your class as the sample, collect data on the number of children in the family and the number of boys. Graph the data to see if there is a positive correlation between the number of boys and the size of the family.

1.5 Exploring Linear Inequalities

The **trichotomy** rule states that when comparing two values a and b, only three situations can occur: $a = b$, $a < b$, or $a > b$. In the previous section we examined the first case, where two values were equal to each other, called an equation. Now we will consider the other two cases, called **inequalities**. Since we will be examining inequalities that either graph as a line or whose boundary is a linear function, we will be dealing with **linear inequalities**.

GRAPHING LINEAR INEQUALITIES IN ONE VARIABLE

A linear inequality in one variable is an inequality statement containing one variable raised to the first power. (*Note*: Remember, $x^1 = x$.) For example, $2x - 3 < 4$ is a linear inequality in one variable. The inequality $x^2 - 3x + 1 > 0$ is not a linear inequality in one variable; even though this example has only one variable, x is raised to the second power.

Inequalities may be expressed as strictly greater or less, or in combination with the equality statement. For example, $3x + 5 \geq 12$ represents the statement "$3x + 5$ is greater than or equal to 12." In all there are four conditions, represented in the following manner:

$3x + 5 < 12$ strictly less than,
$3x + 5 > 12$ strictly greater than,
$3x + 5 \leq 12$ less than or equal to,
$3x + 5 \geq 12$ greater than or equal to.

The algebraic solution to this type of linear inequality is done in the same manner as solving an equation, except for two cases: If multiplying or dividing by a negative value on both sides of the inequality, reverse the direction of the inequality sign.

EXPLORATION 1: Find the solution set (i.e. values of x that make the inequality true) for the inequality $-5x + 4 < 19$.

We find

$$-5x + 4 < 19$$
$$-5x < 15 \quad \text{(add -4 to both sides)}$$
$$x > -3 \quad \text{(divide by -5 and reverse the inequality)}$$

The solution set is the set of values for x such that $x > -3$. Check the solution by selecting a value in the solution set and substitute it into the original inequality. For example, let $x = 2$:

$$-5(2) + 4 < 19 \quad \text{(test for a true statement)}$$
$$-10 + 4 < 19$$
$$-6 < 19 \quad \textbf{(True} \text{ statement!)}$$

A graph of the solution set for a linear inequality in one variable is a line graph representing the values contained in the set on a number line. The graph for the previous example would be a line graph representing all values greater than -3. Figure 1.72 shows this type of line graph. ◊

Notice that there is an open circle representing the first value on the graph. This identifies a strict inequality where the starting value, -3 in this case, is not included in the solution set. If the original problem had been stated as $-5x + 4 \leq 19$, then the solution set would be $x \geq -3$ and the first point on the graph would be included in the solution set (see Fig. 1.73).

We can draw the graph of these linear inequalities in one variable on the TI-81. Enter the inequality just as given in the example on the Y= menu. The inequality symbols are found on the TEST menu (key 2nd TEST). Figure 1.74 shows this menu. Select the appropriate symbol for the problem; use option **5:** < for this example. Figure 1.75 shows the Y= menu with the inequality entered.

Set the RANGE of the calculator for the basic friendly window, and graph the inequality. Figure 1.76 shows the resulting graph. Why does the graph appear in this

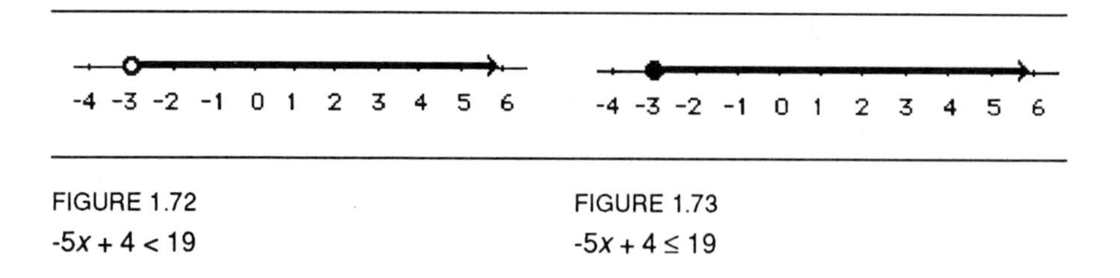

FIGURE 1.72
$-5x + 4 < 19$

FIGURE 1.73
$-5x + 4 \leq 19$

FIGURE 1.74

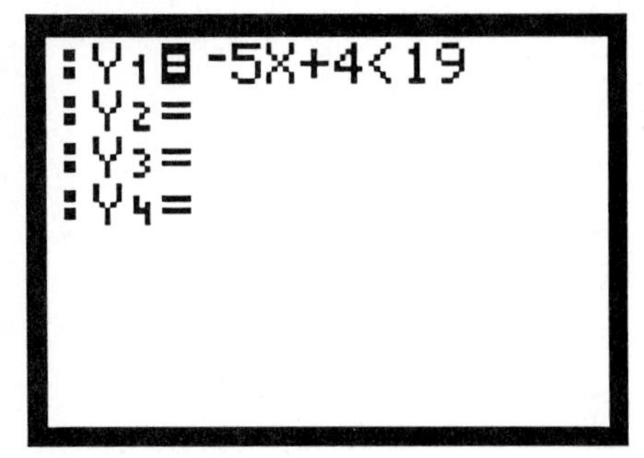

FIGURE 1.75

manner? Change to **Dot** on the **Mode menu** and draw the graph again. What happened? Figure 1.77 shows the graph in **Dot** mode.

Use the TRACE cursor to investigate the values represented on your graph. When $x > -3$, what is the value of y? When $x < -3$, what is the value of y? How do these values compare with the solution set we calculated algebraically? How does the graph on the TI-81 compare with the line graphs shown in Figs. 1.72 and 1.73? Is the endpoint of the solution set, -3, included in the graph?

When we checked the example previously, we chose a value in the solution set and tested the truth of the inequality statement by substitution. Trace to the point **X** = **2**. What is the **Y** value at that point? Why? (See Fig. 1.78.) When substituting values, we are not seeking a numerical "answer," we are only testing for the truth of the statement. Select a point out of the solution set, say -4. By substitution, $-5(-4) + 4 = 24$, so the inequality $-5(-4) + 4 < 19$ is not true. Trace to the point

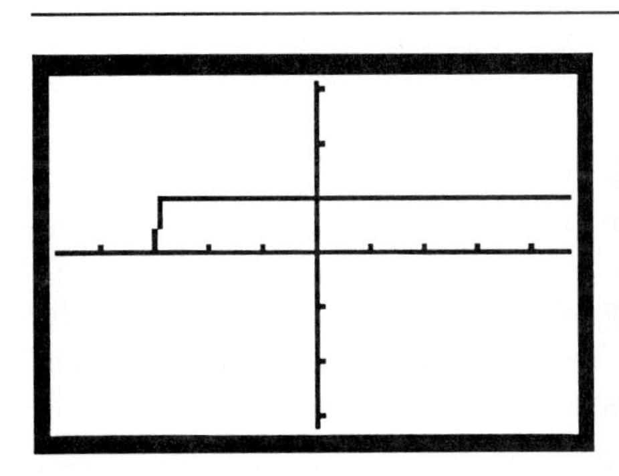

FIGURE 1.76
Connected node

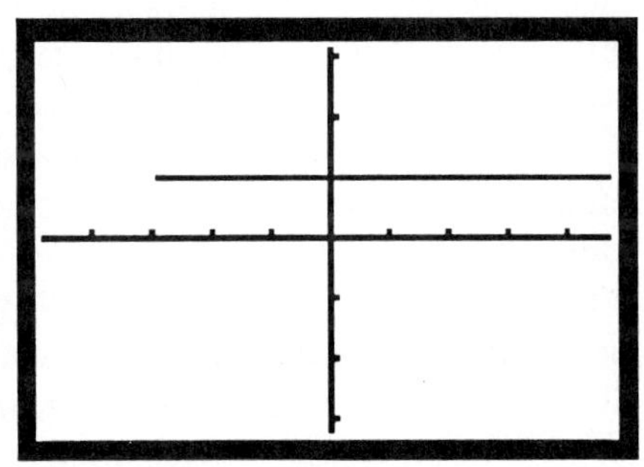

FIGURE 1.77
Dot mode

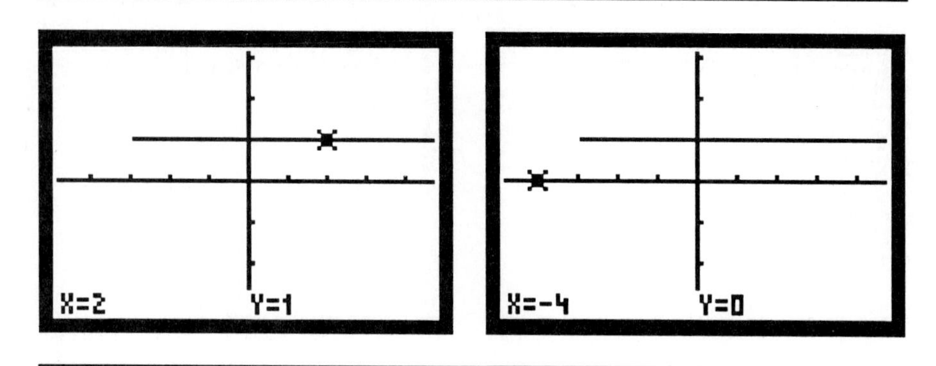

FIGURE 1.78

FIGURE 1.79

X = -4 on the graph (see Fig. 1.79). Where is the TRACE cursor? What is the value of y? Why?

When the selected x value is in the solution set, the function $\mathbf{Y_1}$ has a value of 1, and when the selected x value is not in the solution set, the function has a value of 0. Mathematically we have created a truth table, which returns a value of 1 when the statement is true and a 0 when the statement is false. This is also called a **Boolean relationship,** after the mathematician George Boole who first defined these relationships. A topic for further study could be a report about Boole and the system he created, called Boolean Algebra.

EXPLORATION 2: Solve the following inequality graphically, and then check your solution set algebraically:

$$6x - 2(7 + 2x) \leq -2(1 + 2x).$$

Enter the inequality on the Y= menu (see Fig. 1.80). (Note the difference between the (-) negative key and the subtraction key.) Set the RANGE to the basic friendly window, and draw the graph. Using the TRACE cursor, define the solution set of the inequality. Figure 1.81 shows the graph with the TRACE cursor at the point $x = 2$. Is this point included in the solution set? Why?

Figure 1.81 shows the solution set of all values such that $x \leq 2$. Trace to the left to the point $x = -30$. Does this change your idea of the solution set? Reset the **Xmin = -200**. Is the solution set the same? Try other values for **Xmin:** -1000, -5000, -10,000; does the solution set change? No matter how small you set the value of **Xmin,** the values are still part of the solution set. The set of numbers such that $x \leq 2$ means all values from 2 to negative infinity are included.

Solving the inequality algebraically yields

$6x - 2(7 + 2x) \leq -2(1 + 2x)$

$6x - 14 - 4x \leq -2 - 4x$ (expand by the distributive prop.)

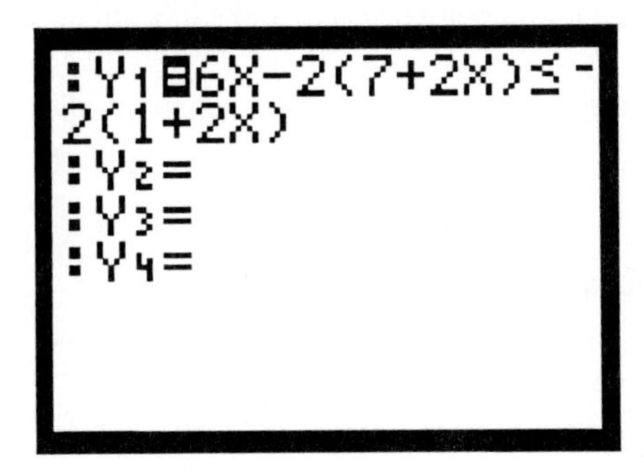

FIGURE 1.80

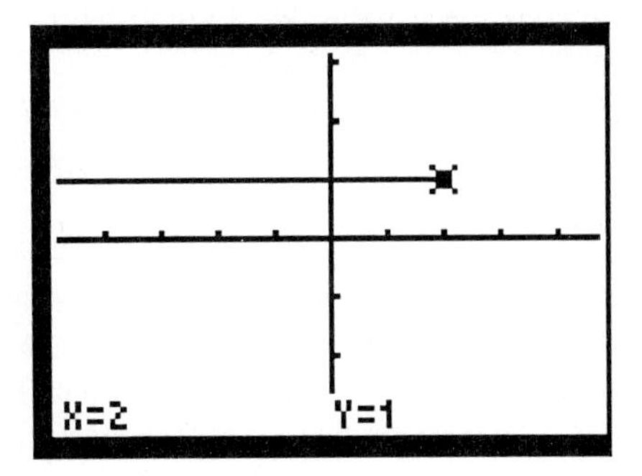

FIGURE 1.81

$$
\begin{aligned}
2x - 14 &\le -2 - 4x && \text{(combine like terms)}\\
6x - 14 &\le -2 && \text{(add } 4x \text{ to both sides)}\\
6x &\le 12 && \text{(add 14 to both sides)}\\
x &\le 2 && \text{(divide both sides by 6)}
\end{aligned}
$$

The algebraic result is the same as the graphical result. Check by selecting any value less than 2 and substituting it into the original inequality. Let $x = 1$; then

$$
\begin{aligned}
6(1) - 2(7 + 2(1)) &\le -2(1 + 2(1))\\
6 - 18 &\le -6\\
-12 &\le -6 && \text{True statement}
\end{aligned}
$$

If we let $x = 2$, we can show that the inequality is still true since the statement says "less than or equal to":

$$
\begin{aligned}
6(2) - 2(7 + 2(2)) &\le -2(1 + 2(2))\\
12 - 22 &\le -10\\
-10 &\le -10 && \text{True statement.}
\end{aligned}
$$

◊

EXPLORATION 3: Another way to express a linear inequality in one variable is a **compound inequality**, or an inequality in three parts as in the following example:

$$-4 < 3x + 2 \le 11.$$

Investigate this inequality.

Read this inequality statement starting in the middle: "$3x$ plus 2 is greater than -4 and $3x + 2$ is less than or equal to 11." The middle term lies numerically between the two outer terms. Correct mathematical notation is to place the smaller value to the left and the larger value to the right. While the expression "$11 \ge 3x + 2 > -4$" is basically correct, order on the number line dictates that the smaller values are to the left and the larger values are to the right.

To graph a three-part inequality on the TI-81, model the statement of the inequality: "$3x$ plus 2 is greater than -4 and $3x + 2$ is less than or equal to 11." Enter

$$\mathbf{Y_1 = (3X + 2 > -4)(3X + 2 \le 11).}$$

The order of each element can vary as long as the mathematical statement is correct. This statement produces a correct graph because the operation between the two terms is multiplication. When both terms are true, they both return a value of 1 and then their product is 1. When either term is false, the product of the terms is 0. An equivalent statement of the inequality would be

$$\mathbf{Y_1 = (-4 < 3X + 2)(3X + 2 \le 11).}$$

Figure 1.82 shows the resulting graph in the basic friendly window. Explore the solution set with the [TRACE] cursor. Because the left side of the inequality is "strictly less than," the left endpoint of the line graph is not part of the graph. In this case, values of $x > -2$ are in the solution set. Figure 1.83 shows the [TRACE] cursor at the right end of the solution set. The point $x = 3$ is part of the solution since the right inequality symbol is "≤." The solution set is $-2 < x \le 3$.

To solve a three-part inequality algebraically, operate in the same manner as with other two-part inequalities, but do the operations to all three parts:

$-4 < 3x + 2 \le 11$

$-6 < 3x \le 9$ (subtract 2 from all three parts)

$-2 < x \le 3$ (divide all parts by 3) ◊

Another type of compound inequality is one whose solution set lies below and above a span of values. For example,

$$2x - 4 < -7.4 \quad \text{or} \quad 2x - 4 > 1.$$

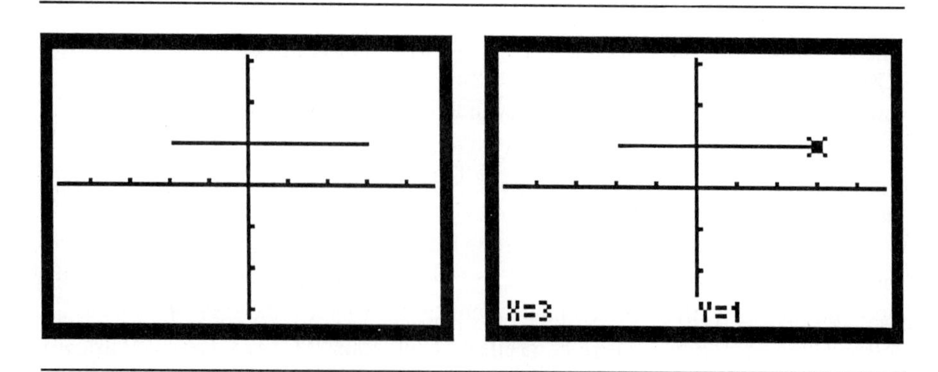

FIGURE 1.82 FIGURE 1.83

We do not state these two conditions as one compound statement because that would imply that the solution set lies somewhere between the values -7.4 and 1. Actually, the solution set is represented as two separate sets of values. To graph the solution set, enter the two separate conditions with a "+" sign between them:

$Y_1 = (2X - 4 < -7.4) + (2X - 4 > 1)$.

The "+" sign returns a true statement (i.e. function value of 1) when either the first or the second statement is true. Figure 1.84 shows the graph drawn in the basic friendly window. Use the TRACE cursor to explore the solution set (see Fig. 1.85). Estimate the solution set from the graph. Are the endpoints of the line segments included in the solution set?

To solve these inequalities algebraically, solve each piece as we have done before:

$2x - 4 < -7.4$	$2x - 4 > 1$
$2x < -3.4$	$2x > 5$
$x < -1.7$	$x > 2.5$

The solution set is $x < -1.7$ and $x > 2.5$. Is this similar to your graphical solution?

GRAPHING LINEAR INEQUALITIES IN TWO VARIABLES

A linear equation in two variables is of the form "$y = mx + b$" (slope-intercept form) or "$ax + by = c$" (standard form). In the previous section we graphed these equations on the TI-81 by entering the equation on the Y= menu. Linear inequalities in two variables are of the form "$ax + by < c$" or "$y > mx + b$," similar to the linear equations.

When we graph linear inequalities in one variable, the solution set is a set of points on the number line. When we graph a linear inequality in two variables, the solution set is the set of points that make up an area above or below a boundary line. The equation of the boundary line between two areas that could be the solution

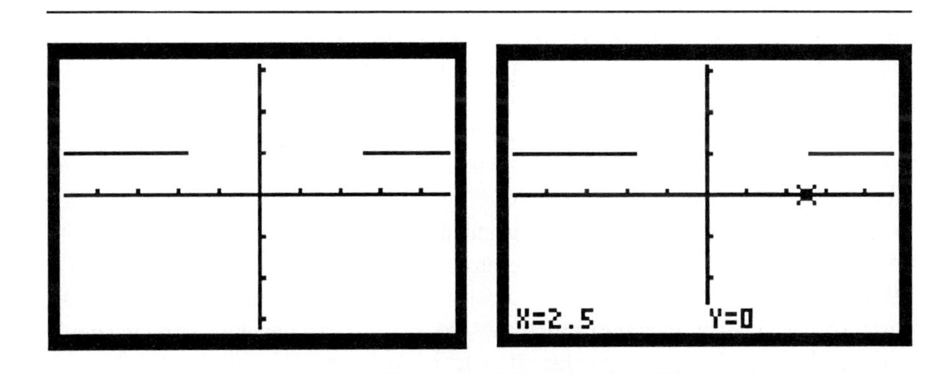

FIGURE 1.84 FIGURE 1.85

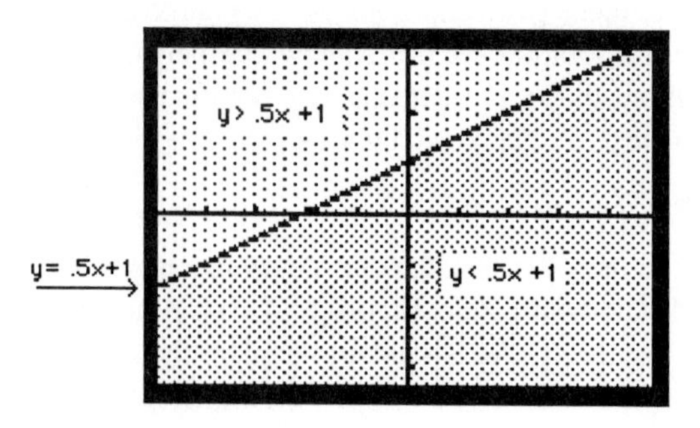

FIGURE 1.86
Three areas of the graph

of an inequality is related to the inequality. Figure 1.86 shows a graph of the line $y = .5x + 1$. Any point in the area above the line is in the solution set of the inequality $y > .5x + 1$, and any point below the line is in the solution set of the inequality $y < .5x + 1$.

If the line $y = .5x + 1$ itself is to be included in the solution set, then the symbols "<" and ">" are replaced with the symbols "≤" and "≥", respectively. To draw the graph of a linear inequality in two variables, first draw the graph of the linear equation, and then shade the appropriate side of the line. Once the equation is in a form that can be graphed on the TI-81 (i.e. slope-intercept), then we will make a decision about which side of the line to shade. This decision is based on whether we want "$y<$" (below the boundary) or "$y>$" (above the boundary).

EXPLORATION 4: Draw the graph of the inequality $y \leq .5x + 1$, show that points in the shaded region and on the line are in the solution set, and show that points in the other region are not in the solution set.

Graph the companion equality $y = .5x + 1$. This is the boundary line between the two areas of the graph (see Figure 1.87). The original inequality includes the symbol "≤," so the line is part of the solution set.

Since the original inequality specifies y values "less than or equal to," the solution set consists of the area on and **below** the boundary line. Select the **Shade** command from the [DRAW] menu. This command shades the area between two specified boundaries. The upper boundary we wish to shade is the line $y = .5x + 1$. Mathematically there is no lower boundary for this area, since theoretically the plane goes on in all directions to infinity. So we must specify some lower boundary at or below the lower edge of the screen. Figure 1.88 shows the **Shade** command entered on the **home screen.** The lower bound is the value -10, which is below the lower edge of the screen,

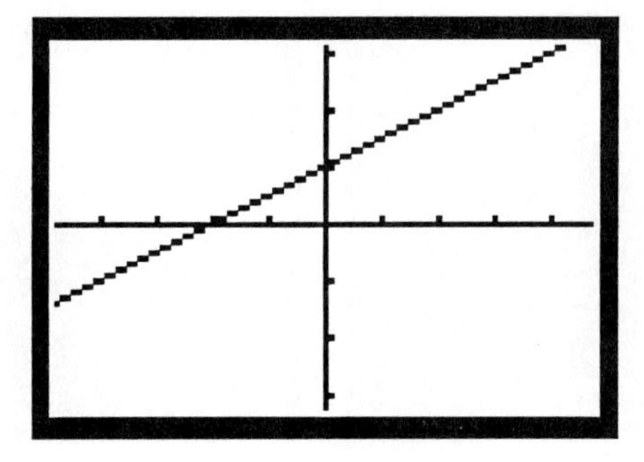

FIGURE 1.87

and the upper bound is the line $y = .5x + 1$. Figure 1.89 shows the resulting graphics screen. ◊

An alternate method for specifying boundaries is to use the RANGE variables **Ymin** or **Ymax** with the inequality. For example the command

Shade (Ymin, .5X + 1)

would shade the same area of the screen as seen in Figure 1.89.

To investigate the points that are in the solution set and those that are not, press any of the four arrow keys to activate the **screen cursor.** Figure 1.90 shows a point in the shaded region (the solution set) and Fig. 1.91 shows a point outside the shaded region (not in the solution set). As you move the cursor around the screen, the x- and y-coordinates of each screen pixel are displayed. These values can be tested to see if they are in the solution set or not. Substitute the coordinates of a point into the original inequality, and test for a true statement.

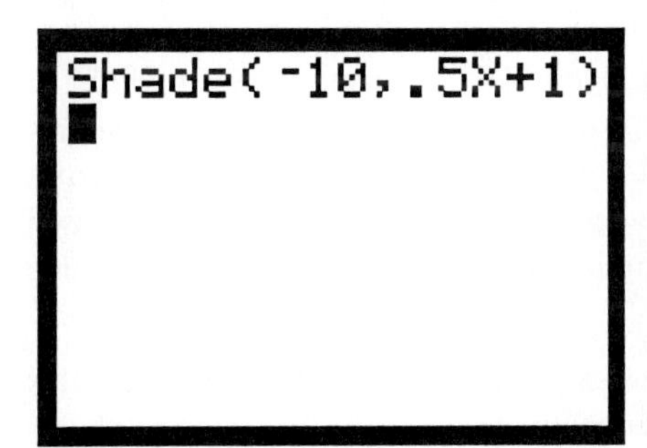

FIGURE 1.88

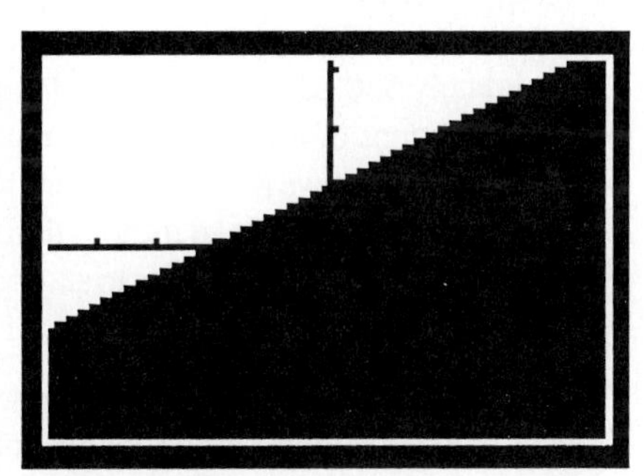

FIGURE 1.89

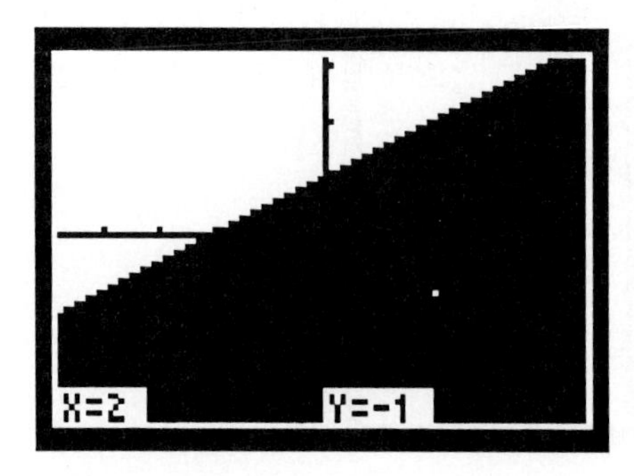

FIGURE 1.90

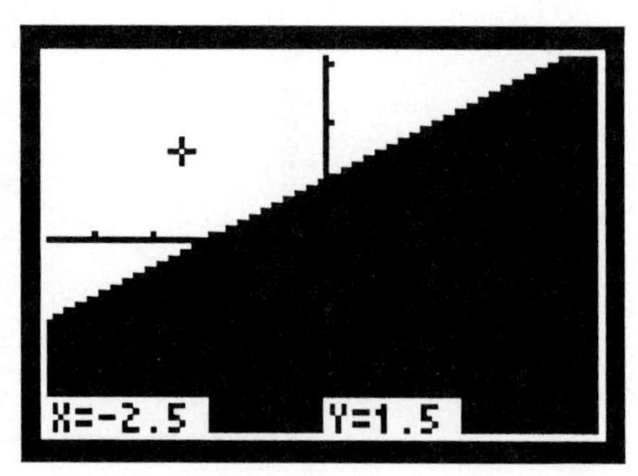

FIGURE 1.91

For example, Fig. 1.90 shows the point (2, -1), which is in the shaded region. Substitute these values into the original inequality:

$-1 \leq .5(2) + 1$
$-1 \leq 1 + 1$
$-1 \leq 2$ **True** statement

Figure 1.91 shows the point (-2.5, 1.5), which is in the unshaded region. Substitute these values into the original inequality:

$1.5 \leq .5(-2.5) + 1$
$1.5 \leq -1.25 + 1$
$1.5 \leq -.25$ **False** statement

Try this for other points in and out of the shaded area. Choose any point on the boundary line. Will these points produce true or false statements? Why? ◊

APPLICATION EXPLORATION: Lloyd Thompson is planning a driving vacation. He has budgeted $1000 to cover travel by car and motel costs for the trip. The cost to operate a car is estimated to be $0.26 per mile driven, and the cost for a motel is approximately $50 per night. Since Lloyd is just planning his trip, he would like to find combinations of miles driven and nights in a motel that will keep him within his budget. Once he knows this information, he can make more specific plans for his trip. Draw a graph that gives Lloyd information about how many days his trip can last based on how far he drives and his desire to stay within his budget. Use the graph to determine the maximum miles Lloyd could drive if he traveled for 7 days.

We will let x represent the number of miles Lloyd drives and y the number of nights he stays in a motel on his trip. Then the inequality

$26x + 50y \leq 1000$

represents the problem situation. We solve this equation for y so we can establish the boundary line of the solution set and decide whether the solution set is the area above or below the line

$$
\begin{aligned}
.26x + 50y &\le 1000 \\
50y &\le -.26x + 1000 \\
y &\le -.0052x + 20
\end{aligned}
$$

Enter the corresponding equation on the Y= menu:

$$y = -.0052x + 20.$$

To set the RANGE we must consider the problem situation. Since x represents the number of miles driven, Xmin can be no smaller than 0. If Lloyd spent all of his budgeted money on the car, then he could travel approximately $\frac{1000}{.26} = 3846.15$ miles. This means that **Xmax = 4000** would be large enough to show the correct graph. The y values represent the number of nights in a motel during the trip. Again **Ymin = 0** (i.e. no nights in a motel) makes sense for the problem. If Lloyd spent all of his travel budget on the motel, he could stay $\frac{1000}{50} = 20$ nights. Let **Ymax = 20**. Set **Xscl** and **Yscl** appropriately and draw the graph (see Fig. 1.92).

Select the **Shade** command from the DRAW menu, and shade the area below the boundary line since we have a "≤" symbol. One version of the command that would shade the correct area is **Shade(0,** $\mathbf{Y_1}$**)**; others will also work. The variable $\mathbf{Y_1}$ in this statement represents the equation stored in position $\mathbf{Y_1}$ on the Y= menu. This variable is found on the **Y-VARS** menu (key 2nd Y–VARS).

Once the area is shaded, activate the **screen cursor** by pressing any of the arrow keys. When the cursor is out of the shaded area, Lloyd is over his budget for that combination of car and motel costs. Figure 1.93 shows the point (2400, 12.063492)

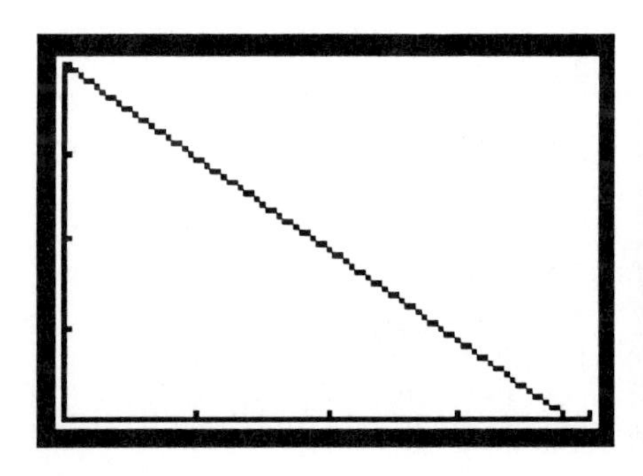

FIGURE 1.92

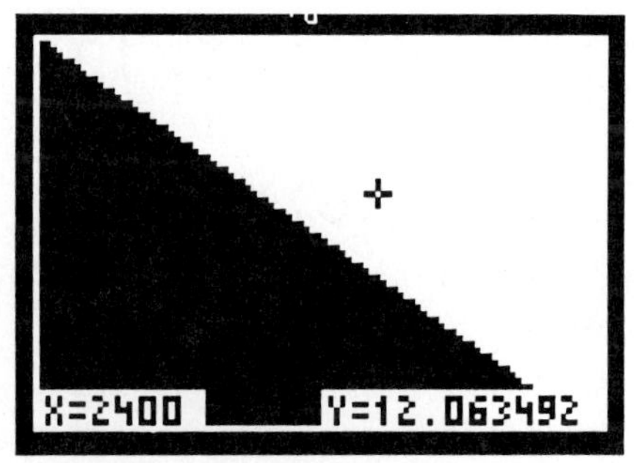

FIGURE 1.93

highlighted. In the problem, this point represents driving 2400 miles and staying in a motel 12 nights. This is more than Lloyd's budget will allow. We show this by substituting this point into the original inequality to check for a true statement (see Fig. 1.94):

$$.26(2400) + 50(12) \leq 1000$$
$$634 + 600 \leq 1000$$
$$1224 \leq 1000 \quad \textbf{False} \text{ statement}$$

Move the **screen cursor** into the shaded region to check a point. Figure 1.95 shows the cursor at the point (1515.7895, 10.15873). This corresponds to Lloyd driving approximately 1516 miles and staying in a motel 10 nights. Substituting into the original inequality gives

$$.26(1516) + 50(10) \leq 1000$$
$$394.16 + 500 \leq 1000$$
$$894.16 \leq 1000 \quad \textbf{True} \text{ statement (see Fig. 1.94)}$$

Notice that for this combination of miles and nights, Lloyd has about \$105.84 left of his \$1000 travel budget (1000 – 894.16 = 105.84). What area of the graph represents the case where Lloyd stays within his budget, but has little or nothing left over? Why?

If Lloyd wants to travel for 7 days, then he will be on the road 6 nights. Move the **screen cursor** to the point where **Y = 6** (approximately). This horizontal line represents all the combinations of miles driven and 6 nights in the motel. The point that represents the most miles driven with 6 nights in the motel while staying within the \$1000 budget is the point along the boundary line. Figure 1.96 shows an estimate of this value: **X = 2694.7368** miles with 6 nights in the motel. If Lloyd drives less than that, he will have money left over; if he drives more than that, he will be over his budget. ◊

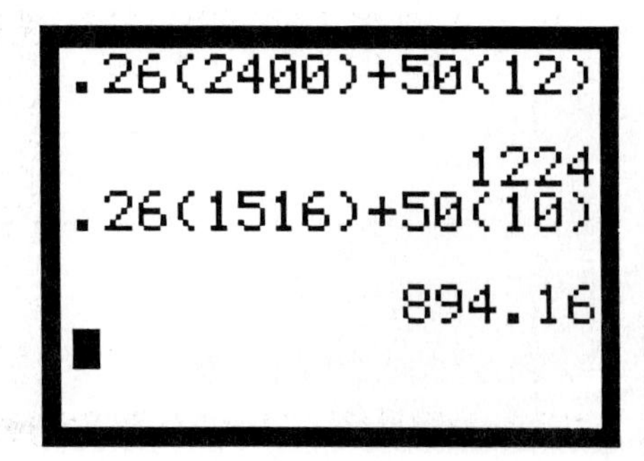

FIGURE 1.94

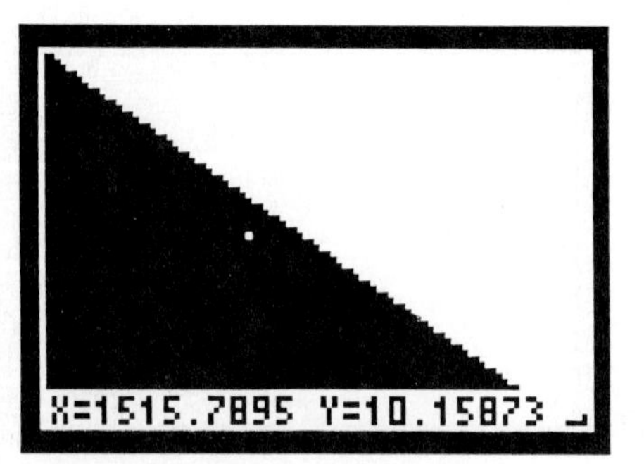

FIGURE 1.95

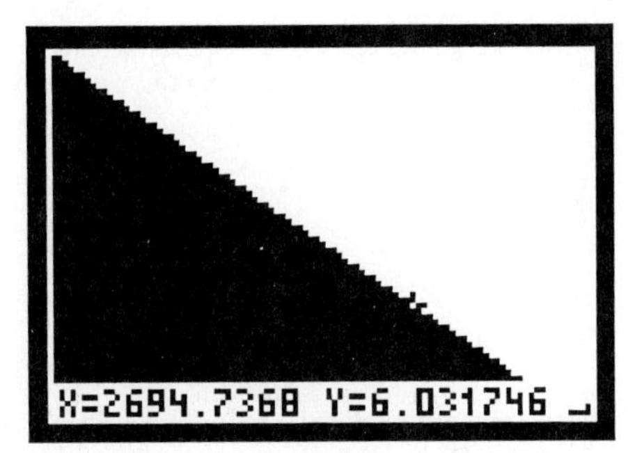

FIGURE 1.96

GRAPHICAL EVIDENCE OF TRUE AND FALSE STATEMENTS

Another way to use the TI-81 to check whether a point on the graphics screen represents a solution to the inequality (true statement) or is not a solution (false statement) is to use a relational operator to evaluate a mathematical statement. As the **screen cursor** moves around the graphics screen, the *x* and *y* coordinates of each point highlighted are stored in the **X** and **Y** memory locations of the calculator. Move the **screen cursor** to the point you wish to check, as in Fig. 1.95. Return to the **home screen** by pressing the [CLEAR] key. Enter the original inequality using the variables **X** and **Y** as seen in Fig. 1.97. If the current values of **X** and **Y** make the statement true (i.e. the point is a solution), then the value of the expression is 1. To evaluate another point, press [GRAPH] and then move the **screen cursor** to another point, such as shown in Figure 1.93, which is not in the solution area. Press

FIGURE 1.97
True statement

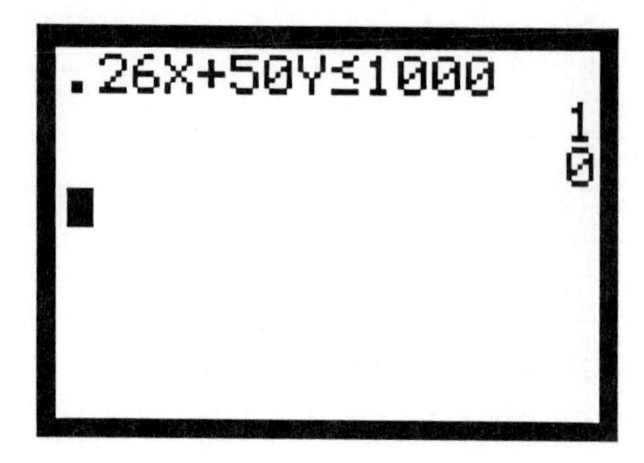

FIGURE 1.98
False statement

CLEAR to return to the **home screen** and then ENTER to evaluate the phrase based on the new values for **X** and **Y**. The value returned by the test statement will be zero if the statement is false (see Fig. 1.98).

This process may be repeated as many times as necessary. Use the TRACE cursor for points on the line in the same way.

Problems

1. Find the solution set of the following inequalities graphically and check algebraically. From the graph choose several points in and out of the solution set, and show that they do or don't satisfy the inequality.

   a. $4x - 3 > 13$ **b.** $3x + 8 < 12$

   c. $5x - 3 < 2x + 1$ **d.** $-7 < x < 3$

   e. $-5 < 2x - 1 < 7$ **f.** $3x + 2 < -7$ and $3x + 2 > 0$

   g. $3x - 4 < 6$ or $2x + 1 > 8$ **h.** $2x > 5$ or $3x > 10$

2. You can also graph an equality in the same manner as in Exercise #1 above, but only a point will be graphed. This can best be accomplished in **Dot** mode, but the solution must correspond to a pixel value of the screen that is set from RANGE . Hence in the basic friendly viewing rectangle, $\mathbf{Y_1 = 3X - 1 = 5}$ will graph a point at $\mathbf{X = 2}$ but $\mathbf{Y_1 = 3X - 1 = 4}$ will not graph a point at $\mathbf{X} = \frac{5}{3}$ because the point 1.666666667 cannot be found on the screen. In fact, if you store $\frac{5}{3}$ in **X** and execute $\mathbf{Y_1}$ from the **home screen**, you will get a 0 for false instead of a 1 for true. Try the following problems and explain why any do not work.

   a. $5x - 3 = 7$ **b.** $10 + 4x = 15$

   c. $3x + 8 = -5$ **d.** $9 - 5x = -12$

In problems 3–6, write an inequality and graph it in a single variable to solve the problem. Check the solution both algebraically and by substituting a value from your solution set into the inequality. Also give the values that make sense for each problem, i.e. whole numbers, integers, rationals, or reals.

3. A car rents for $14.50 per day plus $.11 per mile. How many miles can you drive in a day if you are on a travel expense budget of $100 per day?
4. Car Rental Company A charges $20 a day and $.10 a mile and Car Rental Company B charges $15 a day and $.13 a mile. How many miles must you drive in one day before Car Rental Company A is cheaper than Car Rental Company B?
5. If you were to invest $10,000 at the two different rates of 8% and 11%, then what is the most you can invest at the safer lower rate to earn at least $1000 interest in one year?
6. Susan received $20 for her birthday and wants to buy as many baseball cards as possible along with the display album. If the cards cost $.45 per package of ten and the display album costs $5, what is the most number of cards that she can buy?
7. At what age do we pass a million seconds old? A billion seconds old?
8. Graph the following inequalities in two variables using the **Shade** command. First check algebraically by substituting in points from the shaded region and showing they satisfy the inequality while points from outside the shaded region do not. Then check graphically by using a test statement to show points in the shaded region satisfy the original inequality (true statements) while points from outside the shaded region do not satisfy the original inequality (false statements).

   a. $y < 2x + 3$ **b.** $y > 3 - 5x$

   c. $2x - 3y > 5$ **d.** $-2 < y < 3$
9. For each of the following sets of points find a linear inequality with a solution set containing all of the points. Check your answer using the **Pt-On** command and graphing the inequality in two variables. Answers may vary.

   a. (1, 2), (-2, 3), (0, 5) **b.** (-4, -2), (4, -5), (1, 1)

In problems 10–14, write an inequality and graph it in two variables to solve the problem. Check your solution algebraically and graphically, and state the points you used to check your answer. Give the values that make sense in each problem, i.e. whole numbers, integers, rationals, or reals.

10. A car dealer is willing to sell up to 100 cars and trucks at a discount. Draw a graph showing the different number of cars and trucks that could be sold at a discount.
11. The difference of two numbers is less than 50. Draw a graph showing all of the possible combinations of numbers.
12. Bill Bridger has 100 ft of fence to build a run for his dog. He wants to make a rectangular pen, but is not sure how long to make the length and width. What are some possible lengths and widths of a rectangular pen? What if the pen had to have dimensions greater than 9 by 12 ft?
13. Checking accounts at the Third National Bank have no service charge as long as you maintain a minimum balance of $250. Let **X** represent the amount of deposit or withdrawal to the account and **Y** represent the balance in the account. Make a graph showing that the sum of the deposits or withdrawals and the original balance must be greater than $250. (*Hint:* Positive values of **X** represent a deposit and negative values of **X** represent withdrawals.)

14. Winn High School is playing for the District football championship. Their stadium only holds 1000 people. If the price of tickets is \$2 for children and \$3 for adults, then how many different combinations of tickets could be sold? How much money could be collected?

15. The TRACE cursor can be used as a "counter" as well as for finding specific values of a function. Set your RANGE to the friendly integer viewing window, and graph the following number games using **X** as the original number. Use the TRACE cursor to find the result and to convince yourself that it works for any whole number. Then prove it algebraically. Challenge: Make up your own number game using original patterns.

a. Double any whole number, add nine, add the starting number, divide by three, add four, and subtract the starting number.

b. Add any whole number to the next whole number, add seven, divide by two, and subtract your starting number.

FOR ADVANCED ALGEBRA STUDENTS

16. Graph the following inequalities in one variable, and check algebraically and graphically by choosing points from inside and outside your solution set to test the inequality.

a. $|x-5|<4$ **b.** $|x+2| \geq -3$

c. $x^2+x+1<0$ **d.** $x^3>100$

e. $\frac{1}{x} \leq \frac{3}{1+x}$ **f.** $\sqrt{x+1} > \sqrt{1-x}$

g. $\log x < 10$ **h.** 2^$x < 30$

i. $\sin x \leq 0$ **j.** $5-|x| \geq |x|-5$

17. Graph the following inequalities in two variables, and check by choosing points from in and outside your solution set to test the inequality.

a. $y>x^2$ **b.** $y \leq 1-x^2$

c. $y \leq \sqrt{1-x^2}$ **d.** $y<\sin x$

e. $y<\sqrt{x}$ **f.** $y \geq \cos x$

18. The **Shade** command has several features that you should be aware of. You can start and stop the shading at any x-value you wish. On the home screen enter the command **Shade(Y1,Y2,#,X1,X2)**, where # is the shading factor (# = 1 is the best), X1 is the value to start shading, and X2 is the value to end the shading. Hence the command **Shade(2X, X+1, 2, 3, 5)** will shade the region between the graphs $y_1 = 2x$ and $y_2 = x+1$ in resolution 2 only from $x = 3$ to $x = 5$. Graph the following inequalities for the intervals indicated. Experiment with various resolution settings.

a. $y \leq 2x+1$ from $x=0$ to $x=3$ **b.** $y<1-x$ from $x=-2$ to $x=5$

c. $y>6x-7$ from $x=-5$ to $x=9$ **d.** $y \geq 7-3x$ from $x=0$ to $x=3$

e. $-3<x\leq 4$

19. From the **Standard** viewing window, shade each of the four quadrants separately.

20. If the length of a rectangle is 10 more than its width, then when is the perimeter greater than its area? What if the length is twice the width?

21. When an object is propelled vertically into the air, the height above the ground at any time is given by $H(t) = -16t^2 + v_0 t + s_0$, where t is the time, in seconds, v_0 is the initial velocity of the object, and s_0 is the starting height. Suppose a baseball was thrown vertically from level ground at 75 feet per second.

a. Write and graph an equation using time on the x-axis and height on the y-axis.

b. What values make sense in the problem situation?

c. Use the **Shade** command to show when the ball is at least 50 feet above the ground.

d. When will the ball hit the ground?

e. What is the maximum height the ball reaches and at what time?

22. Using a square and a circle, show graphically when the area of each figure is less than the perimeter and when the area is greater than the perimeter. Why are there no negative solutions? Why are proper fractions (i.e. $0 < \frac{a}{b} < 1$) always part of the solution set?

23. Graphically check out the following inequalities for different values of a, and make conclusions about the truth of the general statements.

a. $|x + a| \leq |x| + |a|$

b. $|x - a| \geq |x| - |a|$

c. $|a||x| < |ax|$

d. $|a||x| > |ax|$

24. In many of the problems that we solve, the values that make sense in the problem situation are positive integers. An interesting function that you should investigate is the **greatest integer function.** This function is written as $y = [x]$, and it is defined as the greatest integer less than or equal to x. From the [Y=] menu select the [MATH] menu, select the **NUM** submenu, and choose the command **INT**. This will write the command in $\mathbf{Y_1}$ on the [Y=] menu. Enter the graphing variable **X** and draw the graph in a friendly viewing window. You will need to be in the **Dot** mode to see the "steps." Now use [TRACE] to convince yourself that you understand how this function works. How many x-intercepts are there? Is this a function? Explain.

25. A parking garage charges $.75 for the first hour or less and $.50 for each additional hour or part of an hour. Using the **INT** function, write and graph the relationship between cost and the number of hours parked. What does the x-intercept mean?

1.6 Exploring Systems of Linear Equations

A **system of linear equations** is a set of two or more linear equations in two variables that represent the same problem situation or that are graphed on the same set of axes.

EXPLORATION 1: Draw the graphs of the system of two linear equations

$y = -2x + 1$ and $y = x - 2$.

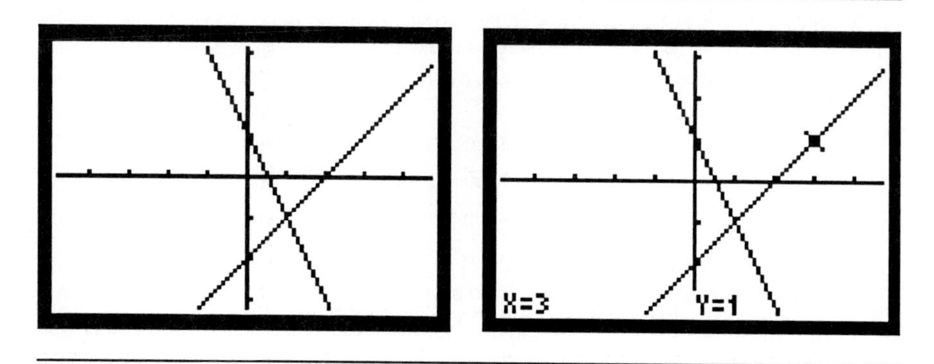

FIGURE 1.99

FIGURE 1.100

Can you predict how the graphs of these two lines will appear before you draw them on the TI-81? Enter the two equations on the [Y=] menu as $\mathbf{Y_1}$ and $\mathbf{Y_2}$. Figure 1.99 shows these two equations graphed with the [RANGE] set to the basic friendly window of [-4.8, 4.7] by [-3.2, 3.1]. Was your guess correct?

Use the [TRACE] function to explore the two lines. Figure 1.100 shows the [TRACE] cursor on the equation $\mathbf{Y_2 = X - 2}$. To switch from tracing on one line to the other, press the [▼] or [▲] arrow keys. The first equation traced is always $\mathbf{Y_1}$. Pressing the [▼] arrow scrolls downward through the [Y=] menu, moving the [TRACE] cursor from one graph to the next. The list of equations is circular in the sense that after the last defined equation is traced, pressing the [▼] arrow causes the [TRACE] cursor to scroll to the top of the list. Pressing the [▲] arrow causes the same process, but in reverse.

What do you notice about the two lines? The line $y = -2x + 1$ has negative slope of -2 and a y-intercept of 1. The line $y = x - 2$ has a positive slope of 1 and a y-intercept of -2. The two lines cross at some point. Estimate the coordinates of this point of intersection from the screen. Activate the [TRACE] cursor to get a better estimate of the point of intersection (see Fig. 1.101). Press the [▼] or [▲] arrow keys. What happened? Why?

It appears that the point (1, -1) is the point of intersection of the two lines. Another way to state this is that the point (1, -1) is on both lines. Let's check this with algebra. If we substitute the point (1, -1) for x and y in both equations, we should get two true statements:

$$(-1) = -2(1) + 1$$
$$-1 = -2 + 1$$
$$-1 = -1 \qquad \textbf{True} \text{ statement}$$

$$(-1) = (1) - 2$$
$$-1 = -1 \quad \textbf{True} \text{ statement}$$

To demonstrate that this point of intersection is special, let's pick a point that is on one line but not on the other and check for two true statements (use the [TRACE] cursor to find points). Use the point (3, 1) which is on the line $y = x - 2$:

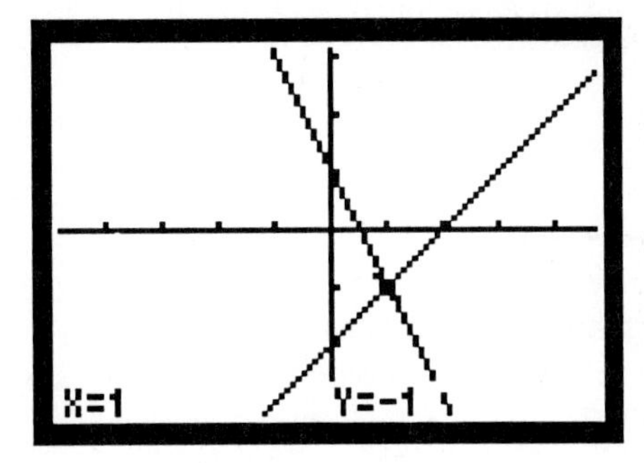

FIGURE 1.101
Point of intersection is (1, -1)

$(1) = -2(3) + 1$		$(1) = (3) - 2$	
$1 = -6 + 1$		$1 = 1$	**True** statement
$1 = -5$	**False** statement		

Try other points on either line to see if you get two true statements when you substitute the same x and y values into both equations. The point of intersection is special!

The solution set for a system of linear equations is the set of points that the lines have in common, or the intersections of the lines. The solution set for this system of equations contains only the point (1, -1). How many points could two lines have in common? ◊

Often, the exact solution to a system of linear equations cannot be found with the original graph as we did in Exploration 1. In these cases we must change the viewing rectangle to get a closer look at the point of intersection. The following exploration demonstrates this procedure.

EXPLORATION 2: Using a graph, find the solution to the following system of linear equations:

$3x + y = -17.37$ and $6x - 5y = -67.5$.

First solve the equations for y (slope-intercept form) so they can be entered into the TI-81.

$3x + y = -17.37$	$6x - 5y = -67.5$
$y = -3x - 17.37$	$-5y = -6x - 67.5$
	$y = 1.2x + 13.5$

Enter the two equations on the [Y=] menu. Let's examine the equations to help set the [RANGE] for the viewing rectangle. The ***x*-intercept** (the point where a linear equation crosses the x-axis) can be found by setting the y value equal to 0 and solving the equation:

$$\begin{aligned} 0 &= -3x - 17.37 \\ 3x &= -17.37 \\ x &= -5.7833 \end{aligned} \qquad \begin{aligned} 0 &= 1.2x + 13.5 \\ -1.2x &= 13.5 \\ x &= -11.25 \end{aligned}$$

The points where the two equations cross the x axis are (-5.7833, 0) and (-11.25, 0). Based on these values, set **Xmin = -12** and **Xmax = 5.**

The y-intercepts of the equations are given by the slope-intercept form of the equations: $y = -17.35$ and $y = 13.5$. Based on these values, set **Ymin = -20** and **Ymax = 20**. Set **Xscl = Yscl = 1**. Figure 1.102 shows the resulting graph, with the TRACE cursor estimating the point of intersection to be approximately (-7.3, 4.7). We cannot use the complete decimal value shown on the screen because the viewing rectangle is too large to read the graph very accurately.

If you substitute the point (-7.3, 4.7) into the original equations, you will get two false statements. Why? We need to get closer to the point of intersection so we can read it more accurately. We will create a sequence of smaller and smaller viewing rectangles around the point of intersection until we can read the point to any degree of accuracy we choose. Based on our first estimate, the x-coordinate is in the interval $\mathbf{-8 < x < -7}$ and the y coordinate is in the interval $\mathbf{4 < y < 5}$. What is the appropriate setting for **Xscl** and **Yscl** for this window?

Reset the RANGE to the following values and draw the graph again.

Xmin = -8,	**Ymin** = 4,
Xmax = -7,	**Ymax** = 5,
Xscl = .1,	**Yscl** = .1.

Figure 1.103 shows the new graph with the TRACE cursor estimating approximately (-7.35, 4.67) for the point of intersection.

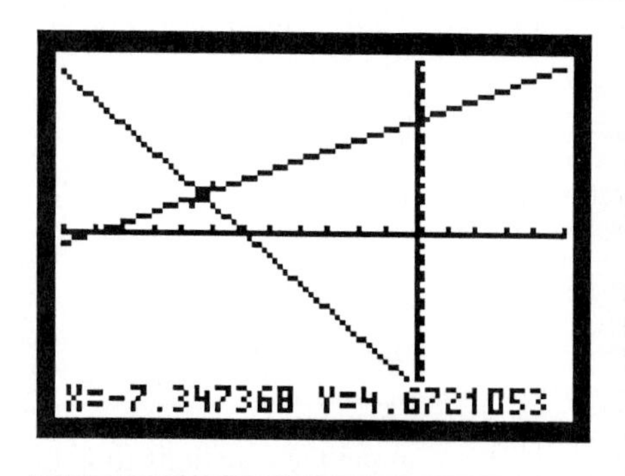

FIGURE 1.102

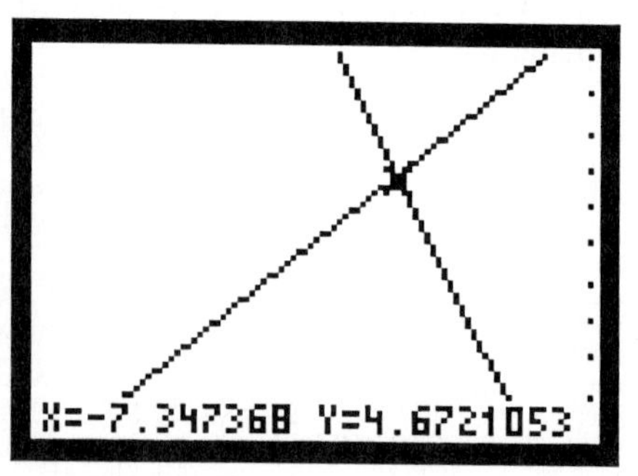

FIGURE 1.103

Based on this estimate, reset the RANGE to the following values and draw the graph:

Xmin = -7.4, **Ymin** = 4.6,
Xmax = -7.3, **Ymax** = 4.7,
Xscl = .01, **Yscl** = .01.

Figure 1.104 shows the new graph with the TRACE cursor estimating the point of intersection to be approximately (-7.351, 4.682). Reset the RANGE to the following values (why?), and redraw the graph (see Fig. 1.105):

Xmin = -7.36, **Ymin** = 4.67,
Xmax = -7.34, **Ymax** = 4.69,
Xscl = .001, **Yscl** = .001.

The new estimate of the point of intersection is (-7.3501, 4.6803).

Reset the RANGE again to the following values and redraw the graph (see Fig. 1.106):

Xmin = -7.351, **Ymin** = 4.679,
Xmax = -7.349, **Ymax** = 4.681,
Xscl = .0001, **Yscl** = .0001.

At this stage we can say that the point of intersection is (-7.350, 4.680) with **error of at most** .002 in both the horizontal and vertical directions. The amount of error is determined by looking at the best known reference points that we have, the RANGE settings: **Xmax – Xmin** = **.002** (-7.349 – (-7.351) = .002) and **Ymax – Ymin** = **.002** (4.681 – 4.679 = .002). If the point of intersection is visible within a viewing rectangle that is at most .002 units across in both directions, then we can be sure that our visual approximation is **at least** this accurate. In fact, any point in the window would be an approximation with this degree of accuracy. This process is called a **zoom in** on the graph. The process is just like zooming in with a video

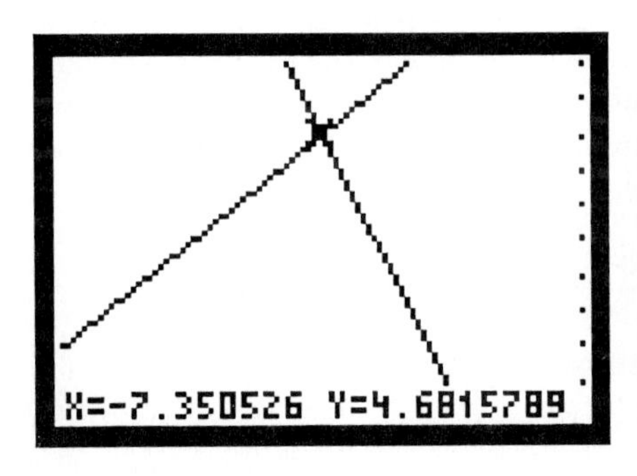

FIGURE 1.104

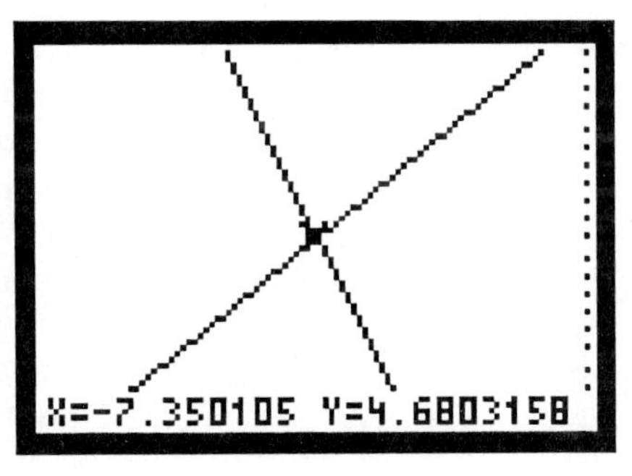

FIGURE 1.105

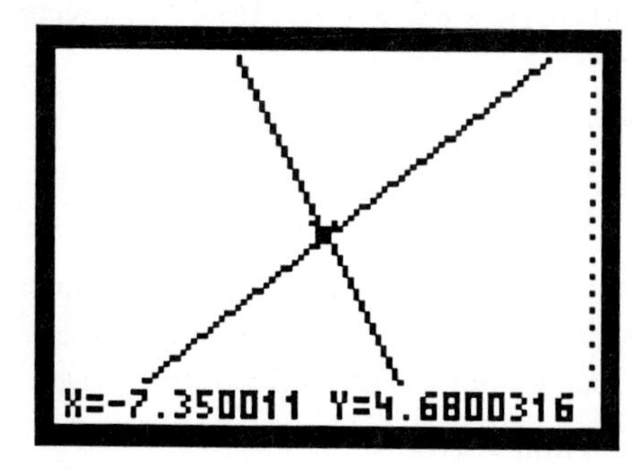

FIGURE 1.106

camera or using a microscope to look at very small areas of the graph. We will use this process many times in future problems.

We could also solve this problem algebraically by setting the slope-intercept forms of the original equations equal to each other and then substituting the x value back into either of the original equations.

If $y = -3x - 17.37$ and $y = 1.2x + 13.5$, then

$$-3x - 17.37 = 1.2x + 13.5$$
$$-4.2x - 17.37 = 13.5$$
$$-4.2x = 30.87$$
$$x = -7.35$$

Substituting $x = -7.35$ into $y = 1.2x + 13.5$ and $y = -3x - 17.37$ gives

$$y = 1.2(-7.35) + 13.5 \qquad y = -3(-7.35) - 17.37$$
$$y = -8.82 + 13.5 \qquad y = 22.05 - 17.37$$
$$y = 4.68 \qquad y = 4.68$$

The solution set to this system of equations is the point (-7.35, 4.68). This is consistent with what we found by zooming in on the graph. Both the algebraic and graphical solutions are correct. Linear equations are special since we can always solve them algebraically. With other equations that cannot be solved algebraically, often the best solution is by zooming in on the graph. ◊

Sometimes special cases of systems of linear equations occur.

EXPLORATION 3: Graph the following system of linear equations, and find the solution set (i.e. the point of intersection):

$$6x + 3y + 12 = 0,$$
$$10x + 5y - 15 = 0.$$

To enter this system of equations into the TI-81, we must solve both equations for y (slope-intercept form):

$$6x + 3y = -12 \qquad\qquad 10x + 5y = 15$$
$$3y = -6x - 12 \qquad\qquad 5y = -10x + 15$$
$$y = -2x - 4 \qquad\qquad y = -2x + 3$$

Enter the two slope-intercept equations on the [Y=] menu. Set the [RANGE] of the viewing rectangle to [-9.6, 9.4] by [-6.4, 6.2], which is twice the basic friendly window. If you are using the ***Friendly Range Program***, use a factor of 2. Figure 1.107 shows the resulting graph. What is the point of intersection between these two lines? Maybe our viewing rectangle is too small to see the intersection. Change the [RANGE] to [-50, 50] by [-50, 50]. Do you see the intersection now? Figure 1.108 shows this viewing rectangle.

How would you describe these two lines? If we examine the two equations in slope-intercept form on the [Y=] menu, you will notice that they both have the same slope but different y-intercepts. Lines that have the same slope but different y-intercepts are **parallel.** Does the graph of these two lines support this conjecture? Can two parallel lines have a point of intersection? Does this system of linear equations have a solution? We say that the solution set for parallel lines is the **empty set**, since there are no points in the set.

Algebraically, since we cannot choose a point that is on both lines at the same time, there is no way to find two true statements. If we set the two equations in slope-intercept form equal to each other, something strange happens. Since $y = -2x + 3$ and $y = -2x - 4$,

$$-2x + 3 = -2x - 4$$
$$3 = -4 \qquad \textbf{False} \text{ statement}$$

We know that $3 \neq -4$, so we say that this is an **inconsistent solution,** meaning that

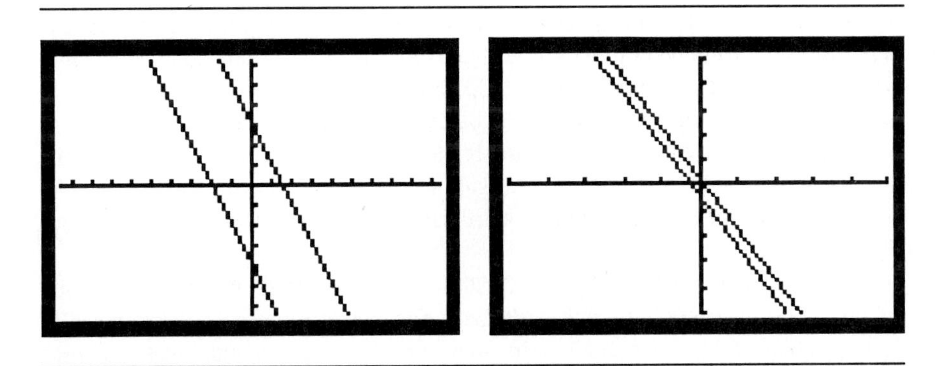

FIGURE 1.107 FIGURE 1.108

the solution is not valid. This confirms what we see in the graph when there are no points of intersection. ◊

EXPLORATION 4: Graph the following system of linear equations and find the solution set (i.e. the point of intersection):

$$6x - 12y = 9,$$
$$2x - 4y = 3.$$

Solve both equations for their slope-intercept form:

$$\begin{aligned} 6x - 12y &= 9 \\ -12y &= -6x + 9 \\ y &= .5x - .75 \end{aligned} \qquad \begin{aligned} 2x - 4y &= 3 \\ -4y &= -2x + 3 \\ y &= .5x - .75 \end{aligned}$$

Notice that both equations reduce to the same slope-intercept form. This means that the system of two equations can be represented as the same line. Do these two lines have any points in common? If you graph both lines, they coincide (lie on top of one another). Try it! Every point on one line is a point on the other line. There are an infinite number of points in the solution set. This is called a **dependent system,** because the two lines coincide.

APPLICATION EXPLORATION: The Shelby Manufacturing Co. produces mountain bikes, which sell for $149.95. Each bicycle costs $110.50 to manufacture. The factory has a fixed overhead cost of $120,000 per year. Write algebraic representations for the **cost** of producing bicycles and the **revenue** from selling the bicycles. Let the horizontal axis represent the number of mountain bikes and the vertical axis represent the amount of money. Draw a graph of the two equations representing this problem situation in an appropriate viewing rectangle. Use the graph to estimate the number of bikes that must be sold to cover all production costs (i.e. break even).

Let x represent the number of mountain bikes. Manufacturing costs for one year would be $110.50 for each bike produced plus $120,000 for overhead. As an equation, we could write $y = 110.50x + 120{,}000$, where y represents the total production costs and depends on x, the number of bikes produced. Notice that this is a linear equation with slope of 110.50 and a y-intercept of 120,000.

The revenue gained from selling bikes is $149.95 for each bike sold, or $y = 149.95x$. In this equation, y represents the total revenue and depends on x, the number of bikes sold. Again, this equation is a linear function with a slope of 149.95 and a y-intercept of 0 (why?). Enter both of these equations on the [Y=] menu (see Fig. 1.109).

Setting an appropriate [RANGE] for the graphics screen can best be done by analyzing the nature of the two equations. The x-axis represents the number of mountain bikes sold, so we should start from zero. Since the company makes a profit of about $40 per bike ($149.95 – 110.50 = 39.45), each 1000 bikes sold results in about

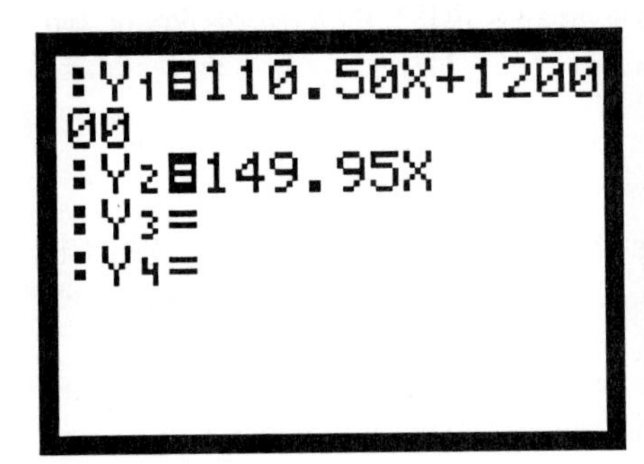

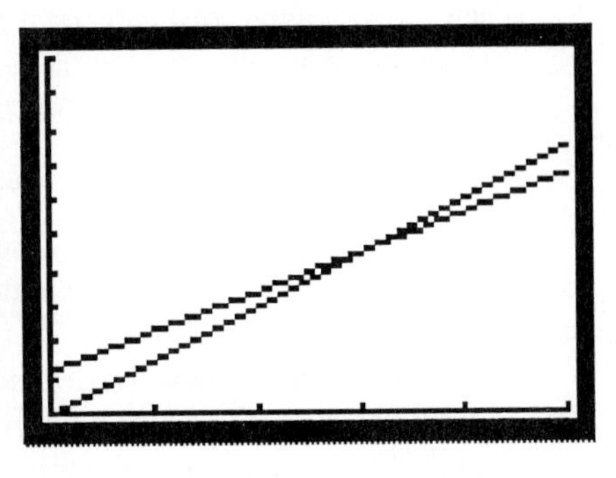

FIGURE 1.109

FIGURE 1.110

\$40 × 1000 = \$40,000 profit. It will take sales of at least 3000 to account for just the overhead (fixed) costs. Let's try a **Xmax** = 5000. **Xscl** values appropriate for this window could be 500 or 1000. The y-axis represents money. Revenue starts at 0 and must be at least \$120,000 to cover overhead costs plus \$110.50 for each bike manufactured. After some experimentation, a **Ymax** = 1,000,000 gives an appropriate viewing rectangle for the problem situation. Figure 1.110 shows the graphics screen with the graphs of the two linear equations drawn in the viewing rectangle [0, 5000] by [0, 1,000,000].

To break even on costs, the amount spent on production ($y = 110.50x + 120{,}000$) must equal the amount received as revenue ($y = 149.95x$). This is the point where these two graphs intersect. Use the TRACE function to estimate the break-even point. (This is shown in Fig. 1.111 in the viewing rectangle [0, 7000] by [0, 1,000,000].)

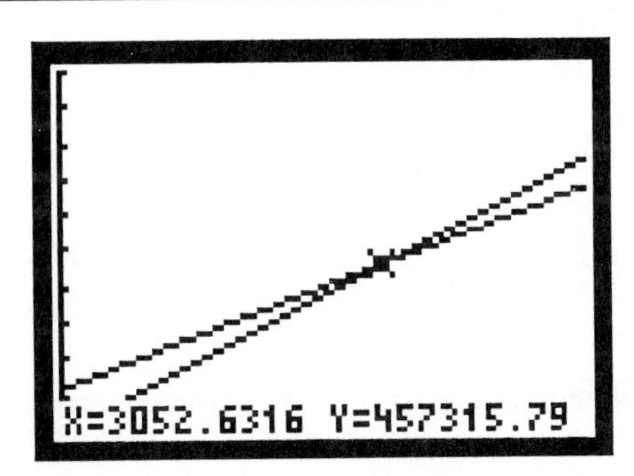

FIGURE 1.111

When x, the number of bikes, is approximately 3052, the cost of production, \$457,315, is approximately equal to the revenue, \$457,742. To confirm this estimate, evaluate the two equations using $x = 3052$:

$$110.50(3052) + 120{,}000 = 457{,}246 \text{ and } 149.95(3052) = 457{,}647.40.$$

(*Note*: These values do not exactly match the values given by the [TRACE] cursor seen in Fig. 1.112 because we rounded 3052.6316 to 3052, since we cannot talk about a part of a bicycle.)

To evaluate a function at a given value quickly on the TI-81, first enter the value 3052 in the **X** memory storage location. Any numeric value stored in **X** will be used in the equations stored in $\mathbf{Y_1}$ or $\mathbf{Y_2}$, or any of the variables on the [Y–VARS] menu. Once a value is stored in **X**, select $\mathbf{Y_1}$ from the [Y–VARS] menu and press the [ENTER] key. The screen will display the value of the function for the given **X** value. To evaluate $\mathbf{Y_2}$ for the same value of **X**, select $\mathbf{Y_2}$ from the [Y–VARS] menu and press [ENTER]. Figure 1.112 shows the **home screen** with these operations and answers displayed. The memory assignment arrow (→) is displayed by the [STO ▶] key. The values are the same as the result previously given.

To get a better estimate of the break even point, change the range to make a window closer to the point of intersection of these two graphs. Some experimentation should lead to a window of approximately [3030, 3060] by [455,000, 458,000]. Use the [TRACE] function to estimate 3042 bikes sold for a break even point of \$456,141 (see Fig. 1.113).

Algebraic solutions for this problem can be found by setting the two equations equal to each other and then solving for x:

$$\begin{aligned} 110.50x + 120{,}000 &= 149.95x \\ 120{,}000 &= 39.45x \\ x &= 3041.825095 \end{aligned}$$

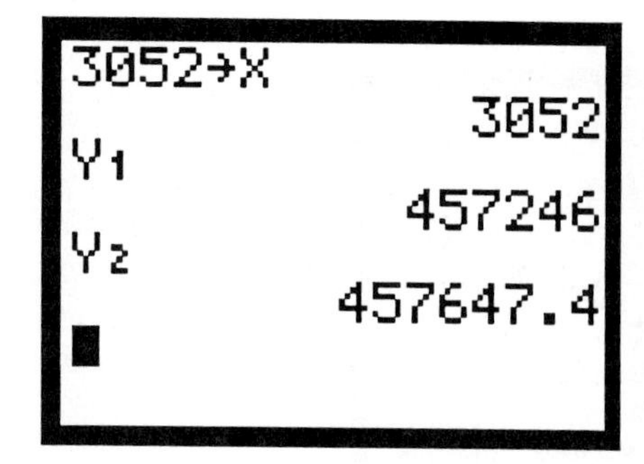

FIGURE 1.112

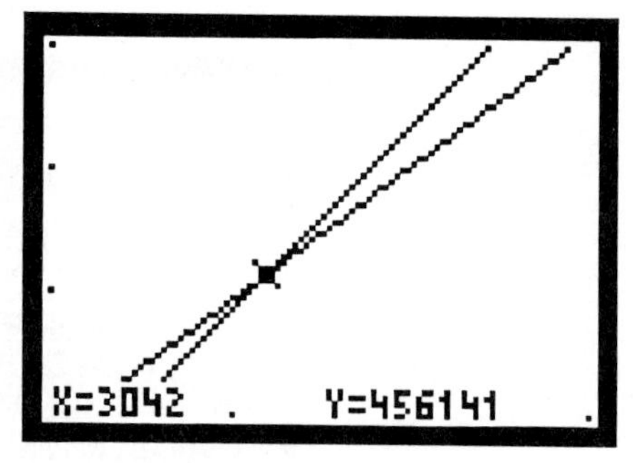

FIGURE 1.113
Break even point

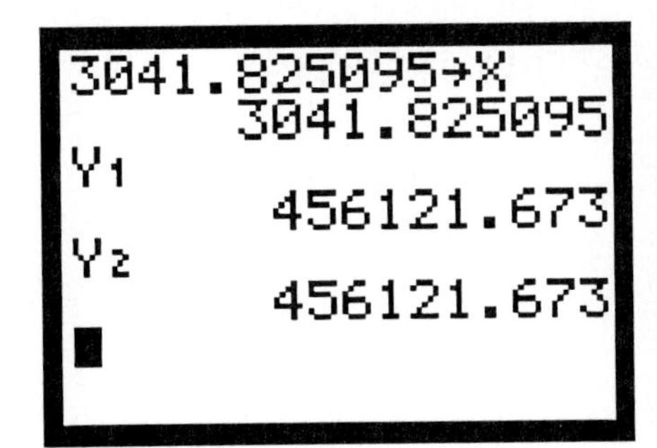

FIGURE 1.114
Functions evaluated

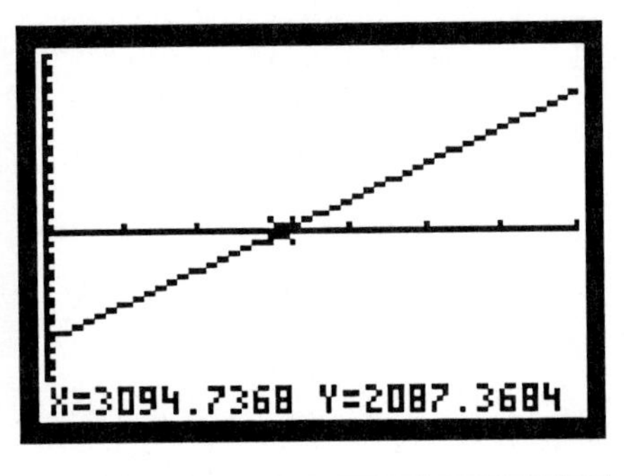

FIGURE 1.115

By substituting this value for x in both equations we get the same y value (see Fig. 1.114):

$$110.50(3041.825095) + 120000 = 456121.673$$

and

$$149.95(3041.825095) = 456121.673$$

For purposes of this problem, the number of mountain bikes must be a whole number. Rounding gives an answer that 3042 bikes must be sold to break even. This is the same answer found by changing the viewing rectangle and reading from the graph using the TRACE cursor.

Another Solution Technique

An alternate solution technique for this problem is to subtract the cost function from the revenue function to create a profit function (i.e. $\mathbf{Y_3 = 149.95X - (110.50X + 120000)}$, or $\mathbf{Y_3 = Y_2 - Y_1}$). By graphing this profit equation, the break even point is the point where the graph crosses the x-axis (i.e. where profit equals zero). Figure 1.115 shows this graph in the viewing rectangle [0, 7000] by [-200,000, 200,000] with the TRACE cursor estimating the break even point. If you zoom in on this graph, you will get the same estimate for the number of mountain bikes needed to break even (see Fig. 1.116 where the viewing rectangle is [3040, 3050] by [-100, 100]).

Algebraically, this profit equation can be solved by setting it equal to zero and solving for the value of x; we get the value as before for the number of bicycles:

$$149.95x - (110.50x + 120{,}000) = 0$$
$$149.95x - 110.50x - 120{,}000 = 0$$
$$39.45x - 120{,}000 = 0$$

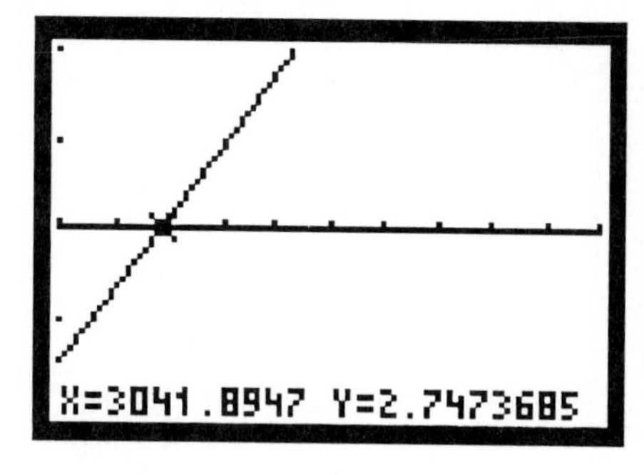

FIGURE 1.116

FIGURE 1.117

$$39.45x = 120{,}000$$
$$x = 3041.825095$$

If you defined the profit equation as $\mathbf{Y_3}$, then by storing **3041.825095 → X**, you can see that this value returns a profit of 0 by evaluating $\mathbf{Y_3}$ (see Fig. 1.117). ◊

Problems

1. Use TRACE to find the x-intercept within .01 for the following linear functions, and check algebraically.

   a. $y = x + 2$ **b.** $y = 3 - 2x$ **c.** $2x + 3y - 1 = 0$

2. Use the **Line** command to draw a line through the following x- and y-intercepts. What is the slope? Now graph the line $\frac{x}{a} + \frac{y}{b}$, where a is the x-intercept and b is the y-intercept. Make a conclusion.

   a. (0, 4) and (5, 0) **b.** (0, -3) and (-1, 0) **c.** (0, 6) and (-5, 0)

For problems 3–8 write an equation, solve it graphically, and check algebraically.

3. When a weight was hooked to a rubber band, it was found to stretch according to the equation $l = 9.2 + 1.7m$, where l represents the length of the stretched rubber band in centimeters and m represents the size of the weight in grams. Graph the equation so that the weight is on the x-axis and the length is on the y-axis.

   a. Find the x- and y-intercepts.

   b. Explain what each intercept means in the problem situation.

   c. Solve for m in terms of l, and graph the new equations so that the length is on the x-axis and the mass is on the y-axis.

   d. Now explain the meaning of the x- and y-intercepts in the problem situation.

4. The cost of renting a car is \$100 plus \$.25 a mile. Suppose that you had a coupon for \$150. Write an equation to show the relationship between the cost and the number of miles you drove. How far could you drive and owe nothing?

5. The cost of renting a car is \$100 plus \$.25 a mile. Write and graph an equation that would show the correct cost for each mile you drove. What meaning does the x-intercept have in this problem?

6. For children under 18 years old the following formula is used to calculate the amount of sleep they need: $H = 17 - \frac{a}{2}$, where H is the hours of sleep and a is the age. Write and graph the equation to calculate hours of sleep needed for any child. When would the child need no sleep? Does this make sense in the problem situation?

7. The speed, S, in words per minute at which a typist can type is calculated using the formula $S = \frac{w}{5} - 2e$, where w is the number of words typed in five minutes and e is the number of errors. Assuming only 5 errors, write and graph an equation showing the relationship between the speed and the number of words typed in five minutes. How many words would a person have to type to have a score of zero?

8. Suppose you have \$20 to spend on tapes. Some tapes cost \$2.10 and others cost \$3.25. How many different combinations of the two types of tapes can you buy?
   - **a.** Write and graph an equation that describes the relationship between the two tapes and the amount you have to spend.
   - **b.** In a friendly viewing screen, use TRACE to find all practical solutions.
   - **c.** Suppose you wanted to spend all of the \$20. Why are some numerical combinations not part of the solution set?
   - **d.** What is the meaning of the x- and y-intercept in this problem?

9. Use TRACE to find the solution to the following systems of equations within .01 and check algebraically:

   a. $y = 2x - 3$, $y = (2/3)x + 2$ **b.** $x + y = 10$, $x - y = 10$ **c.** $2x + 3y = 12$, $3x - 2y = 20$

10. For the lines $y1 = m_1x + b_1$ and $y_2 = m_2x + b_2$, the x-coordinate of the point of intersection is $x = (b_2 - b_1)/(m_2 - m_1)$. Check this for the following equations by graphing and using TRACE. Explain what happens if $m_2 = m_1$.
    - **a.** $y_1 = 2x - 1$ and $y_2 = 3x + 4$
    - **b.** $y_1 = -4x + 6$ and $y_2 = x - 3$
    - **c.** $y_1 = 5x - 2$ and $y_2 = 6x + 5$

11. Check the algebra in the solution to the following equation. See if you can explain the final answer by graphing both sides of the equation on the Y= menu.

    $$6x - 14 = 9x - 21$$
    $$2(3x - 7) = 3(3x - 7)$$
    $$2 = 3$$

    What happens to the two lines? Why? Can you solve algebraically in a different way? Make up another equation in which this happens.

For problems 12–20, write a system of equations, solve them graphically, and check them algebraically.

12. If the sum of two numbers is 111 and their difference is 11, what are the numbers?

13. If you have 20 coins made up of nickels and dimes that total $1.50, how many of each could you have?

14. An apartment charges a deposit and monthly rent for each unit. The first month a person paid $600, and at the end of one year he had paid $6000. How much was the deposit and how much was the monthly rent?

15. Suppose someone invested $5000, part at 8% per annum, and the rest at 10% per annum. If at the end of one year the total interest earned was $410.50, what amount was invested at each rate?

16. The measure of one acute angle of a right triangle is 2.2 times the measure of the other. Find the two angles.

17. Two rectangular rooms have a common wall of 12 feet and their perimeters are in a ratio of 3:2. What are the dimensions of the two rooms?

18. On a trip a person can travel 55 mph on the highway, but due to construction must detour onto secondary roads and only average 40 mph. If the trip is 470 miles and takes 10 hours, how long did the detour take?

19. The distance to a thunderstorm can be calculated using the formula $D = \frac{8t}{25}$, where t is the time in seconds from seeing the lightning to hearing the thunder, and D is the distance to the storm in kilometers (1 kilometer is about .6 miles).

a. Draw the graph of the distance to the storm as a function of time to hear the lightning.

b. How far away is a storm from your friend's house if you counted 5 seconds between the lightning and the thunder and your friend lives a mile away in the opposite direction?

c. Explain the meaning of the x- and y-intercepts for this function.

d. Make a display combining your graph with a numerical chart showing the distance to a thunderstorm for various times between 0 and 30 seconds.

20. Bill and Juan were discussing the number of video games they owned. If Bill gave Juan one video, they would each have the same amount. If Juan gave Bill one video, then Bill would have twice as many as Juan. How many videos does each boy have?

21. There is a big difference between the graphs of two equations intersecting at a point, not intersecting at any point, or intersecting at all points. The first case suggests that the equations are equal for a single value, the second case suggests that the two equations are never equal, and the third case suggests that the equations are equal for all values. The third case, where the two equations are equal for all points, is called an **identity**. The graph of both sides of an identity are identical for all x values.

Assuming that you have a proper range, graph both sides of the following equations separately on the [Y=] menu to see if these equations represent identities. If not, change the equations so they are identities, and check by graphing.

a. $2(x+3) = 2x+3$ **b.** $\frac{3+x}{3} = 1+x$ **c.** $\frac{x}{2} + \frac{1}{3} = \frac{x+1}{5}$

FOR ADVANCED ALGEBRA STUDENTS

22. For two equations written in standard form, $Ax+By=C$ and $Dx+Ey=F$, there is a formula for calculating the point of intersection, or solution. The point of intersection (solution) is given by

$$\left(\frac{CE-BF}{AE-BD}, \frac{AF-CD}{AE-BD}\right) \text{ for } AE-BD \neq 0.$$

Confirm this formula for the following systems of equations.

a. $2x+5y=1$
$3x-y=7$

b. $4x-3y=5$
$3x+7y=10$

c. $8y-9x=21$
$7x+11y=19$

23. We know that the diagonals of a parallelogram bisect each other. Using the following coordinates, confirm that the diagonals bisect by graphing their equations and finding their point of intersection:

A (-2, 6), B (4, 6), C (1, 2), D (-5, 2).

24. We know that the three medians of any triangle are concurrent (intersect at a common point). In fact, if the vertices of a triangle are (x_1, y_1), (x_2, y_2), and (x_3, y_3), then the medians intersect at the point

$$\left(\frac{1}{3}(x_1+x_2+x_3), \frac{1}{3}(y_1+y_2+y_3)\right).$$

For the following triangle, show that this is true graphically:

A(1, 1), B(6, 2), C(5, 4).

Can you show that the altitudes are concurrent?

25. A very old problem is one where 100 animals, made up of three kinds, are to be purchased for $100. The prices of each kind of animal are given, and you want to determine the number of solutions. The problem is that there are two equations and three unknowns. In each case given, eliminate one of the variables and graph the function given by the other two in a friendly viewing window.

I. For costs of $7, $2, $0.20 **II.** For costs of $5, $2, $0.25

a. What values make sense in the problem situation?

b. Is the answer the intersection of the two lines?

c. Use TRACE (or **Grid-On**) to help find the solution.

d. Which values satisfy the problem?

e. What happens if each price is less than or equal to $1?

f. Experiment to find other combinations of costs that have more than two solutions.

26. Write the equation of the line through the point (3, 4) and perpendicular to the line $3x+4y=5$. Then graph both lines and find the point of intersection to see if it fits the following formulas:

$$x=\frac{b^2x_1-aby_1+ac}{a^2+b^2}, \quad y=\frac{a^2y_1-abx_1+bc}{a^2+b^2},$$

where (x_1, y_1) are the coordinates of the point, and $ax+by=c$ is the original line. Also check the distance from the point (3, 4) to the line at the point of intersection using the **distance formula,** $D=\sqrt{(x_1-x_2)^2+(y_1-y_2)^2}$.

SOLUTIONS OF SYSTEMS OF EQUATIONS BY MATRICES

27. Another method of using the TI-81 to solve systems of equations is to use **matrices**. After you select the [MATRIX] key, select the **EDIT** submenu to enter the values in the matrices. First, you must dimension your matrices. In the case of a system of two equations in two unknowns, matrix **[A]** will be a 2×2 and matrix **[B]** will be a 2×1. For the general system of two equations in two unknowns,

$$ax + by = c,$$
$$dx + ey = f,$$

you would enter the values for a, b, d, e in a 2×2 matrix **[A]** and c and f in the 2×1 matrix **[B]**. To see your matrices in correct form, return to the home screen ([2nd] [QUIT]) and type [2nd][[A]][ENTER] and [2nd][[B]][ENTER]. Figure 1.118 shows the screen display of the matrices for the system

$$2x + 3y = 5,$$
$$3x + 2y = 3.$$

In matrix form, this system is made up of **[A] [X] = [B]**, where **[X]** is the 2×1 matrix containing the unknown variables x and y. We need to solve for matrix **[X]** by multiplying on the left by the **[A]**$^{-1}$ (the inverse of **[A]**; use the [X^{-1}] key). Enter the problem as seen in Fig. 1.119. Notice that **[A]**$^{-1}$ is multiplied preceding **[B]** (on the left) and **[A]**$^{-1}$ must be defined.

The solution to the system of equations

$$2x + 3y = 5,$$
$$3x + 2y = 3$$

is $x = -0.2$ and $y = 1.8$. Confirm this solution graphically, algebraically, and numerically by substitution.

Solve the following system of linear equations using the matrix technique just shown. Confirm your solution graphically, algebraically, and numerically.

a. $5x + 3y = 7$
$3x + 4y = 5$

b. $4x - 7y = 9$
$7x - 2y = 8$

c. $5x + 8y = 15$
$5x - 8y = 20$

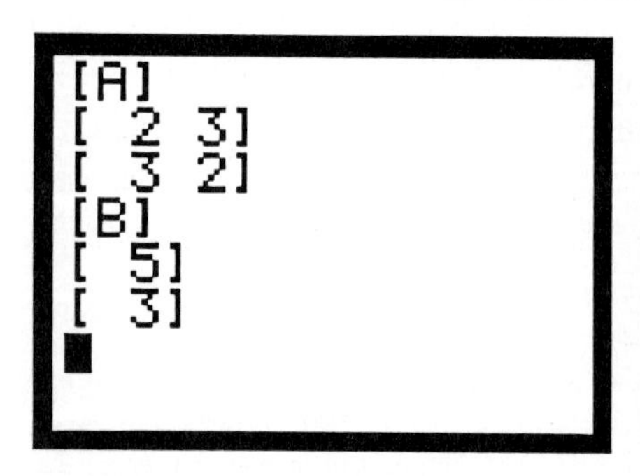

FIGURE 1.118

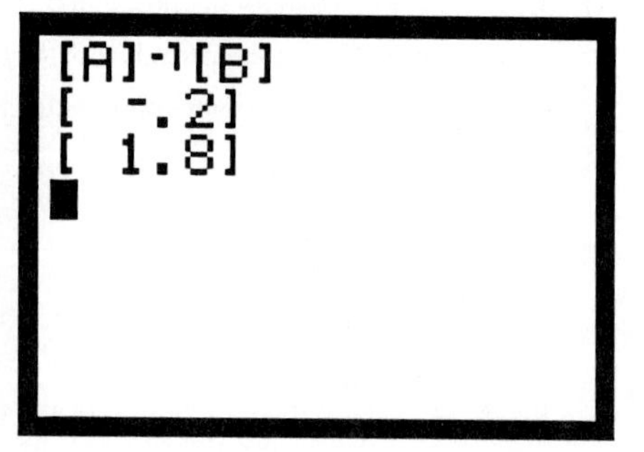

FIGURE 1.119

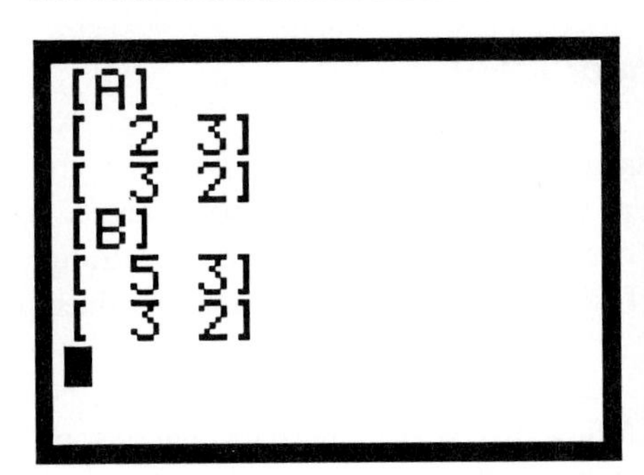

FIGURE 1.120

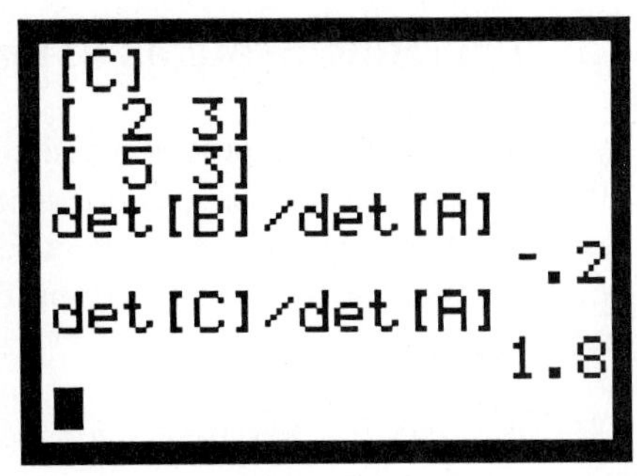

FIGURE 1.121

Cramer's Rule. Another way to use the matrix features of the TI-81 is to use Cramer's Rule, which involves determinants. The determinant of a matrix is calculated using the **5: det** command from the MATRIX menu. Enter the det command before the matrix as shown in Fig. 1.121. For a system of two equations in two unknowns, three 2×2 matrices, [A], [B] and [C], must be created. For the general system

$$ax + by = c,$$
$$dx + ey = f,$$

let $[\mathrm{A}] = \begin{bmatrix} a & b \\ d & e \end{bmatrix}$, $\quad [\mathrm{B}] = \begin{bmatrix} c & b \\ f & e \end{bmatrix}$, $\quad [\mathrm{C}] = \begin{bmatrix} a & c \\ d & f \end{bmatrix}$.

Cramer's Rule states that if det [A] $\neq 0$, then $x = \dfrac{\det [\mathrm{A}]}{\det [\mathrm{B}]}$ and $y = \dfrac{\det [\mathrm{C}]}{\det [\mathrm{A}]}$.

Figures 1.120 and 1.121 show the solution for the previous example done by Cramer's Rule.

Solve problems **(a)**, **(b)**, and **(c)** by Cramer's Rule.

EXPLORING PROBABILITY

28. In a game of chance, the game is considered fair if the **expected value** is zero. The expected value is defined as the probability of winning multiplied by the amount won minus the probability of losing multiplied by the cost of playing, or **Expected Value** = $P(W) \times A - P(L) \times C$, where $P(W)$ = the probability of winning, $P(L)$ = the probability of losing, A = the amount won, and C = cost of playing the game.

a. Write and graph an equation for the expected value as a function of the wager for the following games.

i. Throwing a die and receiving a dollar for a six.

ii. Drawing a card from a standard deck and receiving a dollar for an ace.

b. What is the amount that you need to charge in each game to have a fair game, or expected value of zero?

1.7 Exploring Systems of Linear Inequalities

Systems of linear inequalities are similar to systems of linear equations in the same way as linear inequalities are similar to linear equations. In some cases, a system of linear inequalities is stated as two or more inequalities in the problem setting. In other cases, we can make a system of linear inequalities from a single linear inequality.

EXPLORATION 1: Draw the graph of the following system of linear inequalities and investigate the solution set graphically and numerically:

$x - 2y < 4$ and $x - y > -1$

First solve the inequalities for y:

$x - 2y < 4$	$x - y > -1$
$-2y < -x + 4$	$-y > -x - 1$
$y > 0.5x - 2$	$y < x + 1$

Based on these two equations, we will revise the original question to be: "What is the area **above** the line $y = .5x - 2$ and **below** the line $y = x + 1$?" Can you explain why we used the words "above" and "below" in each case?

Enter the companion equation of the boundary line for each inequality on the [Y=] menu:

$\mathbf{Y_1 = .5X - 2}$ and $\mathbf{Y_2 = X + 1}$

Set the [RANGE] for an appropriate viewing window that shows a **complete graph** of the two lines. A complete graph of the two lines is a graph that shows all the important details of the two lines including x- and y-intercepts and all intersection points. A complete graph in general shows all the important details and behavior of the graphs being drawn. We will discuss this concept again in later sections. Figure 1.122 shows a complete graph of boundary lines of this system of inequalities in the viewing rectangle [-9.6, 9.4] by [-6.4, 6.2].

Shade the area of the graph under investigation. The lower boundary of the system is the line $y = .5x - 2$ and the upper boundary is the line $y = x + 1$. The command **Shade(.5X – 2, X + 1)** or the equivalent statement **Shade($\mathbf{Y_1}$, $\mathbf{Y_2}$)** will shade the area we are investigating. Figure 1.123 shows the shaded area.

Press any of the arrow keys ([◀], [▶], [▼], or [▲]) to activate the **screen cursor**. Move the cursor to the point (3, 2), which is in the shaded region (see Figure 1.124). Points within the shaded region satisfy the conditions of both original inequalities at the same time. Substitute the x and y values of this point into both of the original inequalities:

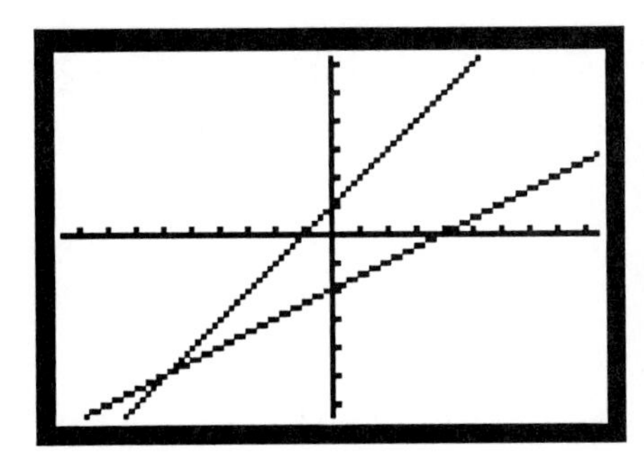

FIGURE 1.122

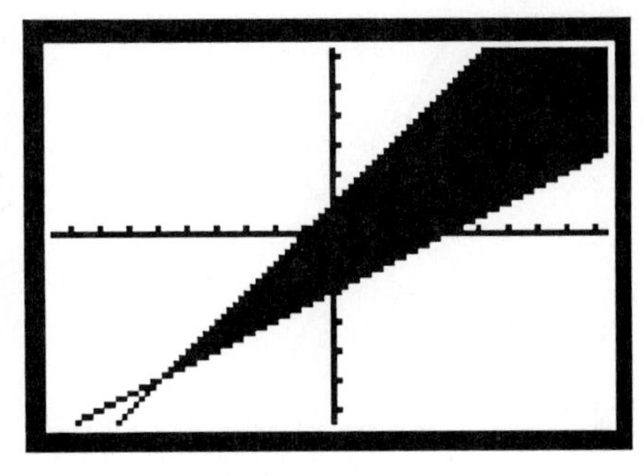

FIGURE 1.123

$$\begin{aligned} x - 2y &< 4 \\ (3) - 2(2) &< 4 \\ 3 - 4 &< 4 \\ -1 &< 4 \quad \textbf{True} \text{ statement} \end{aligned} \qquad \begin{aligned} x - y &> -1 \\ (3) - (2) &> -1 \\ 3 - 2 &> -1 \\ 1 &> -1 \quad \textbf{True} \text{ statement} \end{aligned}$$

Move the **screen cursor** to some point that is not in the shaded region, say (-1.8, 2.4) (see Fig. 1.125). This point is not in the solution set of the system of inequalities. If we substitute the x- and y-coordinates of this point into the original two inequalities, we will **not** get two true statements:

$$\begin{aligned} x - 2y &< 4 \\ (-1.8) - 2(2.4) &< 4 \\ -1.8 - 4.8 &< 4 \\ -6.6 &< 4 \quad \textbf{True} \text{ statement} \end{aligned} \qquad \begin{aligned} x - y &> -1 \\ (-1.8) - (2.4) &> -1 \\ -1.8 - 2.4 &> -1 \\ -4.2 &> -1 \quad \textbf{False} \text{ statement} \end{aligned}$$

In this case, when we evaluated the two inequalities using the point (-1.8, 2.4) we got one true statement and one false statement. Why? Could we choose a point that

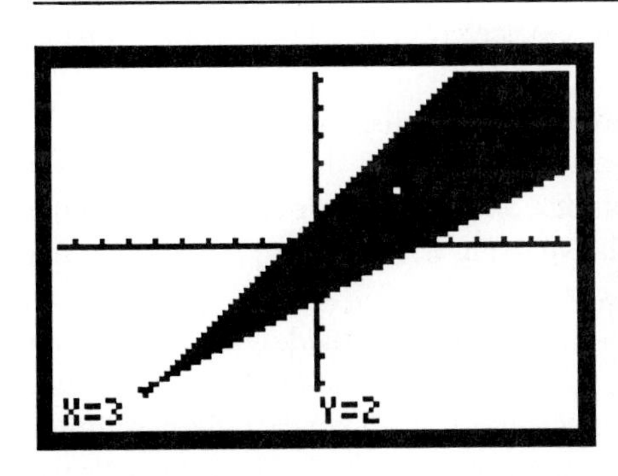

FIGURE 1.124

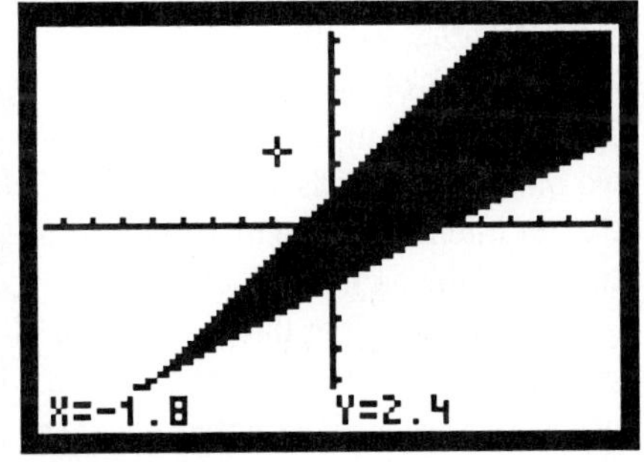

FIGURE 1.125

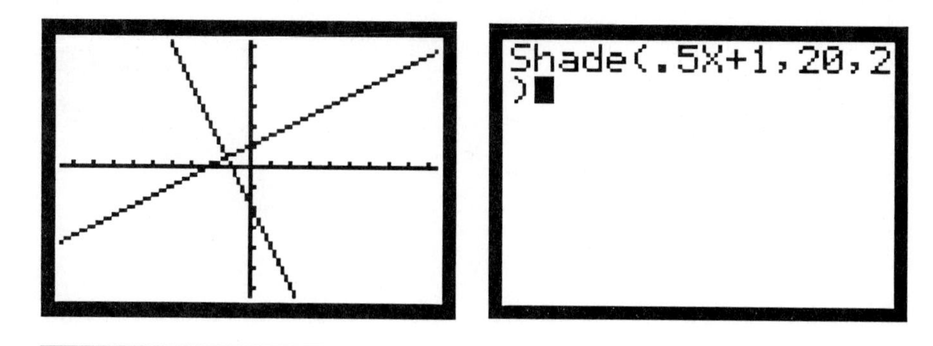

FIGURE 1.126

FIGURE 1.127

would make the first statement false and the second statement true? Could we find one that makes both statements false? Investigate these other conditions and see if you can find the pattern. ◊

In the previous example the region identified as the solution set of the system of linear inequalities was **between** two boundary lines. Sometimes the region we wish to investigate lies **above** two lines. This condition is illustrated in Exploration 2.

EXPLORATION 2: Draw the graph of the following system of linear inequalities and investigate the solution set graphically and numerically:

$$y > .5x + 1,$$
$$y > -2x - 2.$$

Since the two inequalities are in the proper form (i.e. $y > \ldots$), graph the two boundary lines by entering the companion equations

$\mathbf{Y_1 = .5X + 1}$ and $\mathbf{Y_2 = -2X - 2.}$

Figure 1.126 shows the resulting graph on the screen of the TI-81 in the viewing rectangle [-9.6, 9.4] by [-6.4, 6.2].

The area of the graph we wish to investigate lies above the line $y = .5x + 1$ and above the line $y = -2x - 2$ at the same time. To see this area, we will use two **Shade** commands with different resolutions. Select the **Shade** command from the DRAW menu, and enter the parameters shown in Fig. 1.127. Figure 1.128 shows the shading of the graph.

The first entry in the command, **.5X + 1,** is the lower bound of the region. The second entry, **20**, is an upper bound that is off the screen. (You may also use **Ymax** for this entry, since it identifies the top of the screen.) The third entry is the **resolution** of the shading. Resolution can be any whole number from 1 to 8. The value tells the calculator how to choose the columns of pixels on the screen to darken. In Fig. 1.128,

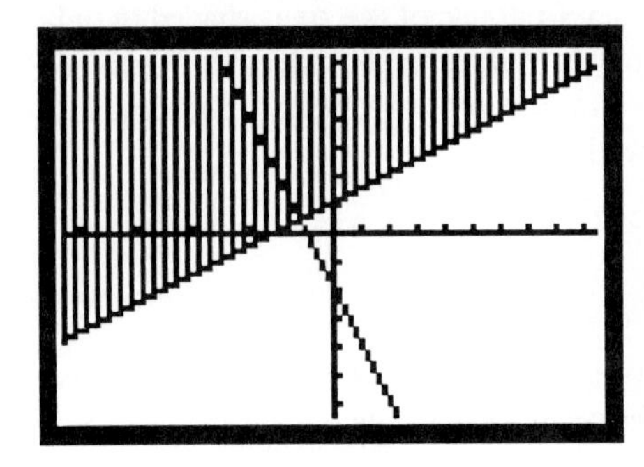

FIGURE 1.128

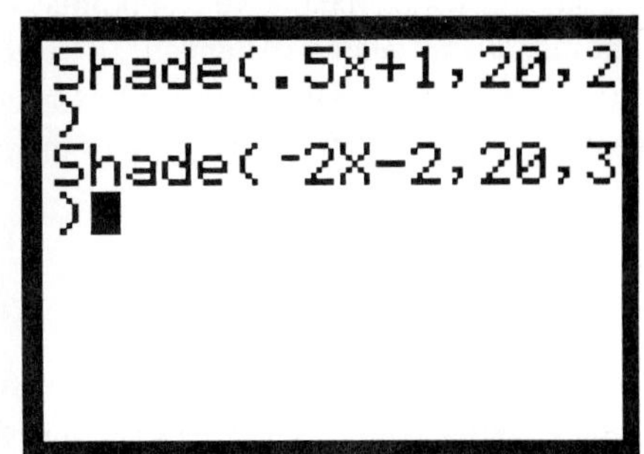

FIGURE 1.129

a resolution of **2** resulted in shading every **second** column of pixels. A resolution of **3** would shade every **third** column of pixels. We will use a resolution of 3 to shade above the line $y = -2x - 2$. Figure 1.129 shows the **Shade** command, and Fig. 1.130 shows the second shading added over the first.

The area of the graph representing the solution set is the area that was shaded by both commands, the area above both lines. Activate the **screen cursor** and move to a point in the double shaded region, for example the point (1, 3). Figure 1.131 shows this point.

By substituting this point into both original inequalities, we will produce two true statements:

$y > .5x + 1$		$y > -2x - 2$	
$(3) > .5(1) + 1$		$(3) > -2(1) - 2$	
$3 > .5 + 1$		$3 > -2 - 2$	
$3 > 1.5$	**True** statement	$3 > -4$	**True** statement

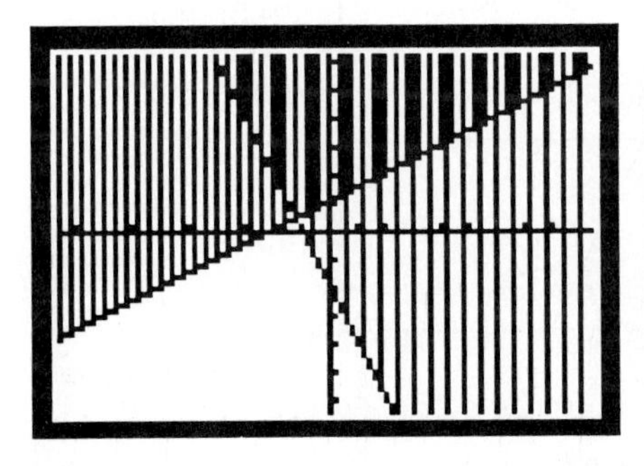

FIGURE 1.130

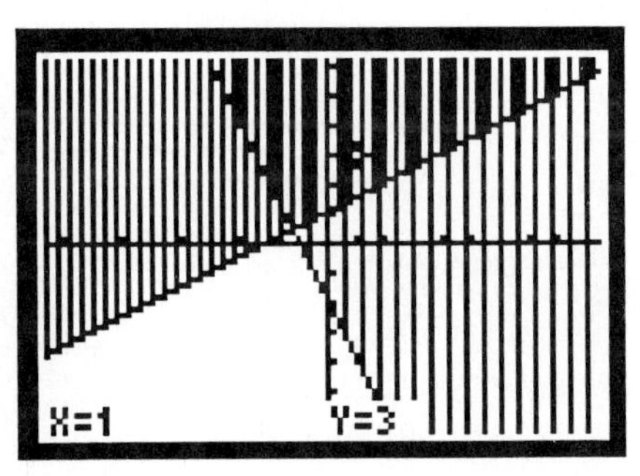

FIGURE 1.131

What will happen if we test points chosen from one of the areas shaded in only one resolution? What will happen if we test points from the unshaded region? Continue to investigate this graph numerically. Can you make a general statement about the four regions created by the intersection of the two boundary lines? ◊

FOR ADVANCED TI-81 USERS

There is a technique that will shade the area above or below two lines with only one **Shade** command. Let's use the system in the previous example:

$y > .5x + 1$ and $y > -2x - 2$.

We entered the boundary lines as $\mathbf{Y_1 = .5X + 1}$ and $\mathbf{Y_2 = -2X - 2}$.

Refer to Fig. 1.132 for the following analysis. The area we wish to shade is above the line $y = .5x + 1$ only when this line is above the line $y = -2x - 2$. And conversely, the area we wish to shade is above the line $y = -2x - 2$ only when this line is above the line $y = .5x + 1$. Make a function statement in $\mathbf{Y_3}$ from this analysis using $\mathbf{Y_1}$ and $\mathbf{Y_2}$ to represent the boundary lines:

$$\mathbf{Y_3 = Y_1(Y_1 \geq Y_2) + Y_2(Y_2 \geq Y_1).}$$

Notice that the symbol "≥" was used in each conditional statement. For the **Shade** command, use $\mathbf{Y_3}$ as the lower bound and either a number above **Ymax** or the command **Ymax** itself for the upper bound. Figure 1.133 shows the command entered on the **home screen**, and Fig. 1.134 shows the resulting shaded area. For some systems

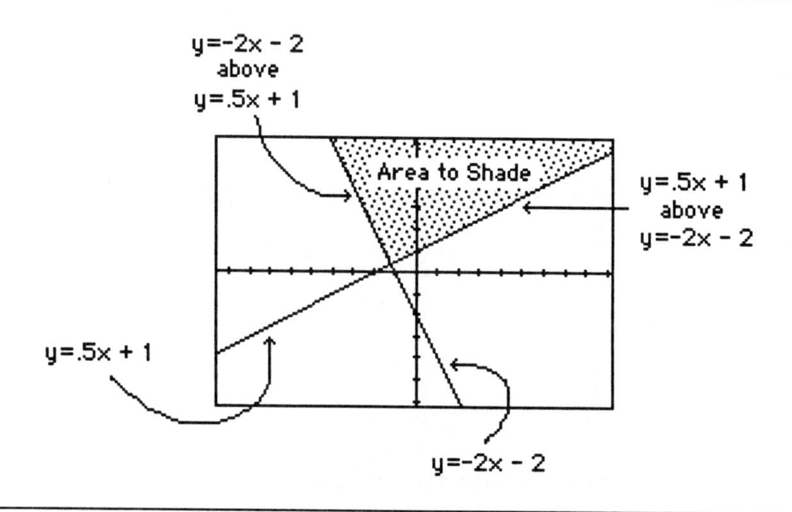

FIGURE 1.132

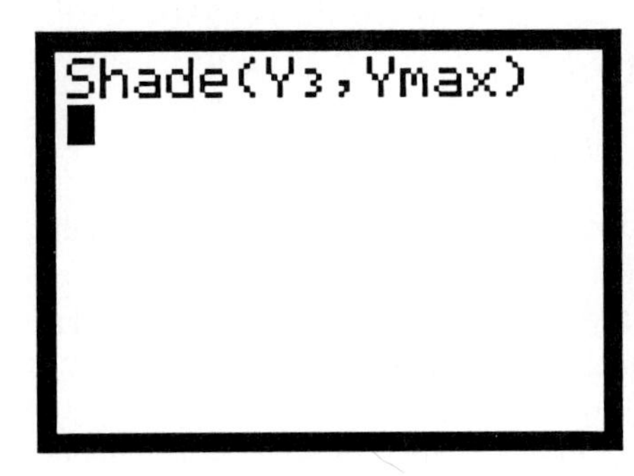

FIGURE 1.133

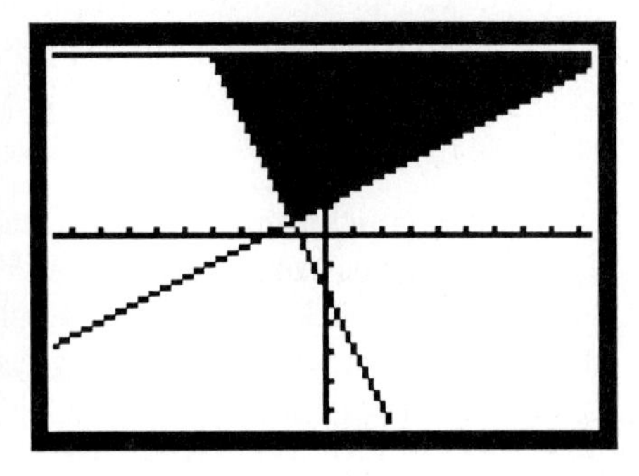

FIGURE 1.134

of equations, a "spike" will be seen near the intersection because of the screen size. Experiment with different combinations of inequality signs (i.e. >, ≥, <, or ≤) in conditional statements like the one shown.

Could you shade the area that lies below both lines at the same time? Experiment!

What would happen if we added a third inequality to the system? Add the inequality $y < .25x + 3$. Enter the boundary line of this inequality as

$$Y_4 = .25X + 3.$$

Since this is now an upper bound (why?), we can shade above both lines $y = .5x + 1$ and $y = -2x - 2$ and below the line $y = .25x + 3$ with the command **Shade (Y_3, Y_4)**. Figure 1.135 shows the shaded triangular region that represents the solution set of the system of three inequalities.

Activate the **screen cursor** and select a point in the shaded region. Substitute this point into all three inequalities, and show numerically that you get three true statements. Use the point (1, 2):

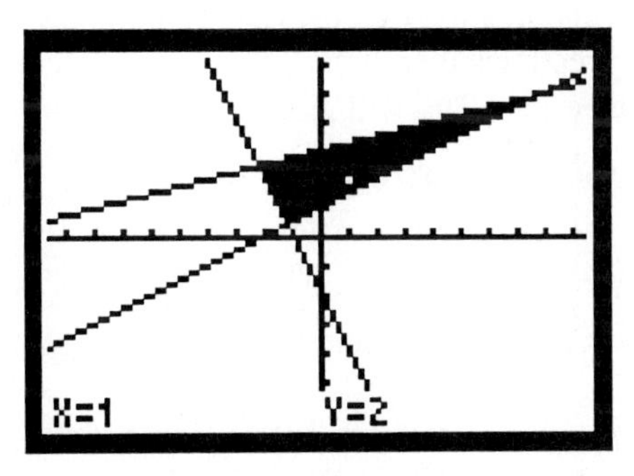

FIGURE 1.135

$y > .5x + 1$	$y > -2x - 2$	$y < .25x + 3$
$(2) > .5(1) + 1$	$(2) > -2(1) - 2$	$(2) < .25(1) + 3$
$2 > .5 + 1$	$(2) > -2 - 2$	$(2) < .25 + 3$
$2 > 1.5$	$2 > -4$	$2 < 3.25$

Try other points in different regions and see if you get three true statements at the same time. Can you make a general rule? Many other combinations of shading under, over, and between lines are possible. Could you incorporate this model for shading into a program that would work for any system of inequalities? Explore!

APPLICATION EXPLORATION: The two most popular cross-training athletic shoes on the market today are *Slammers* and *Jammers*. The Overtime Shoe Stores would like to order at least 800 pairs of the shoes for a summer sale. The wholesale price of *Slammers* is \$80 per pair, and the wholesale price of *Jammers* is \$125 per pair. The accounting department says that the buyer can spend no more than \$80,000 total on the two types of shoes. Draw and shade a graph that represents this problem situation and answer the following questions:

(a) Find different combinations for the order of *Slammers* and *Jammers* which will satisfy the conditions of the problem.

If we let x represent the number of *Slammers* to order and y represent the number of *Jammers* to order, then we know from the problem that

$$x + y \geq 800.$$

Since *Slammers* cost \$80 per pair, *Jammers* cost \$125 per pair, and the total cost of the order cannot be more than \$80,000, we can say that

$$80x + 125y \leq 80{,}000.$$

Solve the two inequalities for the correct form, and enter the companion equality:

$$\mathbf{Y_1 = 800 - X} \quad \text{and} \quad \mathbf{Y_2 = (80{,}000 - 80X)/125.}$$

To set an appropriate **Range** for the viewing rectangle, let's analyze the problem. Since x and y represent the number of each type of shoe to order, both values must be greater than zero. This will give the values **Xmin = 0** and **Ymin = 0**. If we ordered all *Slammers* at \$80 per pair, we could buy $\frac{80{,}000}{80} = 1000$ pairs. If we ordered all *Jammers* at \$120 per pair, we could buy $\frac{80{,}000}{125} = 640$ pairs. Let **Xmax = 1000** and **Ymax = 700**. Figure 1.136 shows the graph of the two boundary lines in the viewing rectangle [0, 1000] by [0, 700].

The area of the graph we wish to investigate is above the line **Y1 = 800 – X** and below the line $\mathbf{Y_2 = (80{,}000 - 80X)/125}$. Why? To shade this area, use either

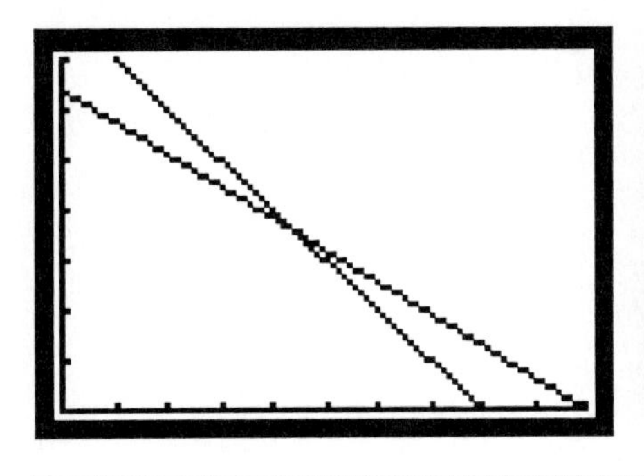

FIGURE 1.136

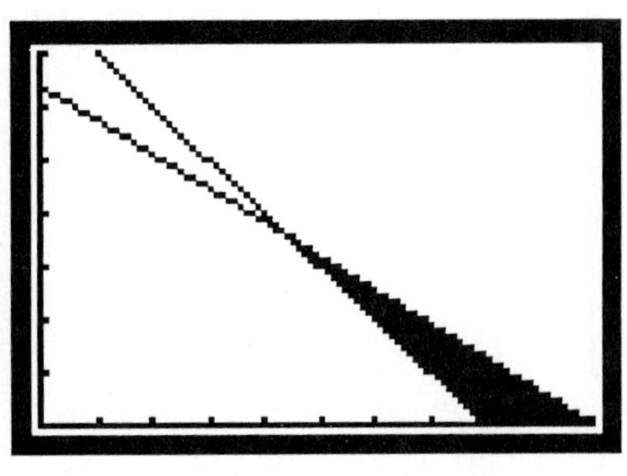

FIGURE 1.137

of the following commands: **Shade(800 – X, (80,000 – 80X)/125)** or **Shade(Y_1, Y_2)**. Figure 1.137 shows the shaded area.

Activate the **screen cursor** and move into the shaded area. Points in the shaded area represent combinations of *Slammers* and *Jammers* that could be ordered within the constraints set by the problem. That is, orders of at least 800 pairs of shoes costing no more than $80,000. Choose a point in the shaded area, and test the x and y values in the original inequalities. Figure 1.138 shows the point (800, 100) highlighted. This represents an order of 800 *Slammers* and 100 *Jammers*.

Substituting, we get

$$x + y \geq 800$$
$$(800) + (100) \geq 800$$
$$900 \geq 800 \quad \textbf{True}$$

$$80x + 125y \leq 80{,}000$$
$$80(800) + 125(100) \leq 80{,}000$$
$$64{,}000 + 12{,}500 \leq 80{,}000$$
$$76{,}500 \leq 80{,}000 \quad \textbf{True}$$

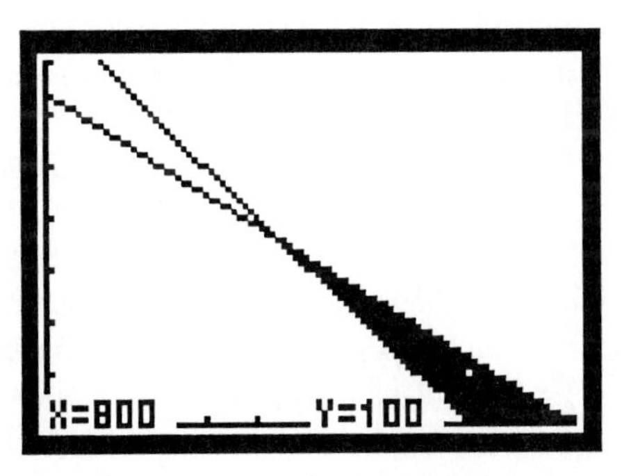

FIGURE 1.138

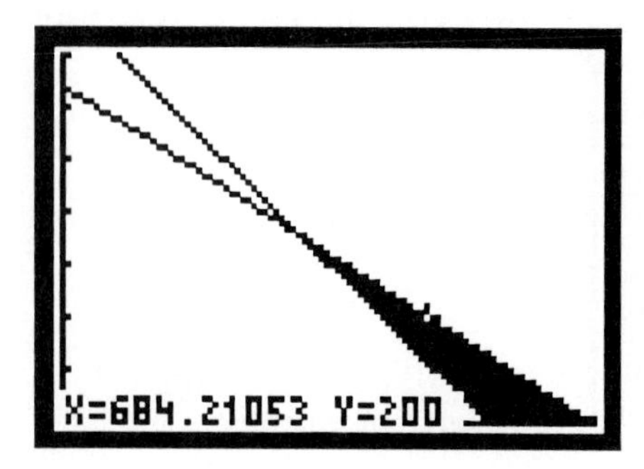

FIGURE 1.139

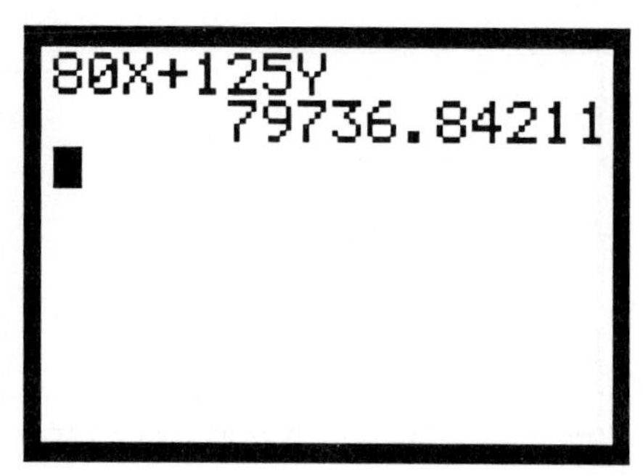

FIGURE 1.140

Any other point in the shaded region will give a similar result.

(b) Which points in the shaded area represent spending the maximum amount of money allowed?

In part (a), notice that the point (800,100) representing 800 pairs of *Slammers* and 100 pairs of *Jammers* cost only $76,500. This means that $3,500 of the $80,000 allowed by the accounting department was not spent. Select other points in the shaded region and calculate the amount of money spent for that order. One way to do this is to move the **screen cursor** to a point in the shaded region. Figure 1.139 shows the point (684.21053, 200), representing an order of 684 pairs of *Slammers* and 200 pairs of *Jammers*. The values shown on the screen for the **X** and **Y** coordinates of the point are automatically stored in the **X** and **Y** memory registers. Press the [CLEAR] key to return to the **home screen**, and evaluate the expression **80X + 125Y** (see Fig. 1.140).

This is a close approximation of the total amount spent on the order represented by the highlighted point. The value is not exact since we must buy whole numbers of pairs. To evaluate another point, press [GRAPH], activate the screen cursor by pressing one of the arrow keys, and move to another point in the shaded area. Once a point is highlighted, return to the **home screen** ([CLEAR] key) and press [ENTER] to evaluate the expression **80X + 125Y** for the new values of **X** and **Y**. Figure 1.141 shows the screen after several points have been evaluated in this way. Notice that the points along the line $\mathbf{Y_2 = (80{,}000 - 80X)/125}$ always represent nearly all of the allowed $80,000 spent on the order. This makes sense based on the conditions of the problem. The upper bound of the problem is the amount of money that can be spent, represented by the inequality $80x + 125y \leq 80{,}000$. The line $\mathbf{Y_2 = (80{,}000 - 80X)/125}$ consists of points where $80x + 125y = 80{,}000$.

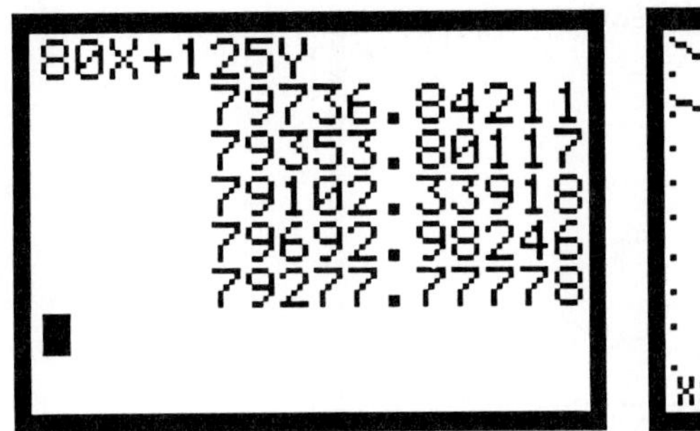

FIGURE 1.141

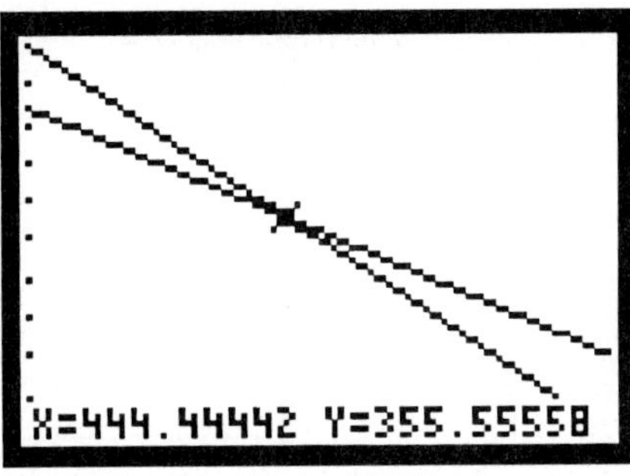

FIGURE 1.142

(c) Which points in the shaded area represent the least amount of money spent while still ordering at least 800 pairs of shoes?

Based on the same reasoning used in part (b), the points in the shaded region that represent the least amount of money spent but still represent at least 800 pairs of shoes are the points in the shaded region along the line $\mathbf{Y_1 = 800 - X}$. Test a few of these points to confirm this conjecture. For example, the point (705.26316, 100) represents a cost of approximately \$68,921.05. This is over \$11,000 less than the allocated amount.

(d) What is the largest number of pairs of shoes which could be purchased? What is the least?

Since *Slammers* are only \$80 per pair, the largest number of pairs of shoes that could be purchased would be an order of all *Slammers*: $\frac{80{,}000}{80} = 1000$ pairs of shoes. At first glance, the least number of pairs that could be ordered would seem to be an order of all *Jammers*. But, an order of all Jammers would be $\frac{80{,}000}{125} =$ 640 pairs of shoes. This contradicts the terms of the original problem, which says that the order must be at least 800 pairs of shoes. This means that the minimum order must be a combination of *Slammers* and *Jammers* that adds up to exactly 800 pairs. There are many combinations satisfying this condition, represented by points in the shaded region along the line $\mathbf{Y_1 = 800 - X}$.

(e) What point represents the most even split between *Slammers* and *Jammers*? Why?

The point of intersection of the lines $\mathbf{Y_1 = 800 - X}$ and $\mathbf{Y_2 = (80{,}000 - 80X)/125}$ represents an order of approximately 800 pairs of shoes costing \$80,000. Find the point by zoom in on the graph. Figure 1.142 shows the viewing rectangle [444.444, 444.445] by [355.555, 355.556] with the TRACE cursor estimating

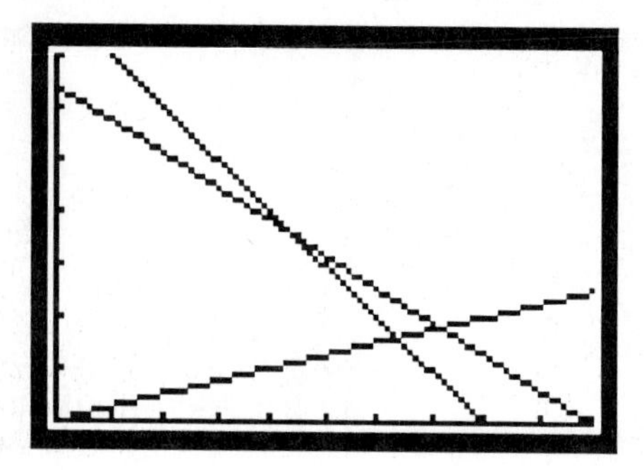

FIGURE 1.143

the point of intersection to be (444.44442, 355.55558) with error of at most 0.01. Why?

This point represents an order of 444 *Slammers* and 355 *Jammers*, costing 80(444) + 125(355) = $79,895. (If we round 355.55558 up to 356, the order is over $80,000!) If we move to other points in the shaded region, the resulting order will be less balanced between the two types of shoes. ◊

PROBLEM EXTENSION FOR ADVANCED ALGEBRA STUDENTS

We will add another constraint to the previous problem. Since we can order many combinations of *Slammers* and *Jammers*, let's specify that the number of *Slammers* ordered must be no more than 4 times the number of *Jammers*. This condition can be

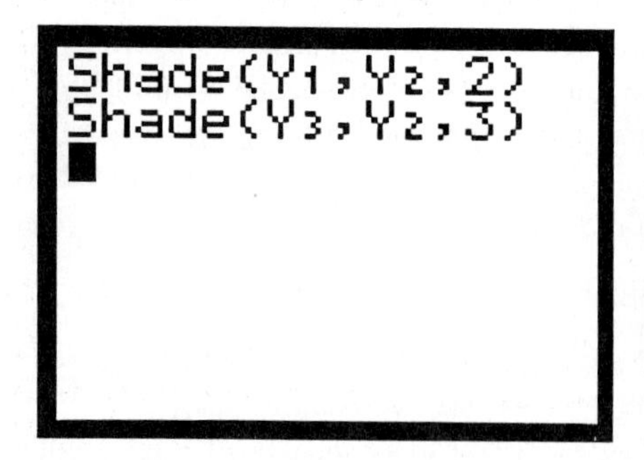

FIGURE 1.144

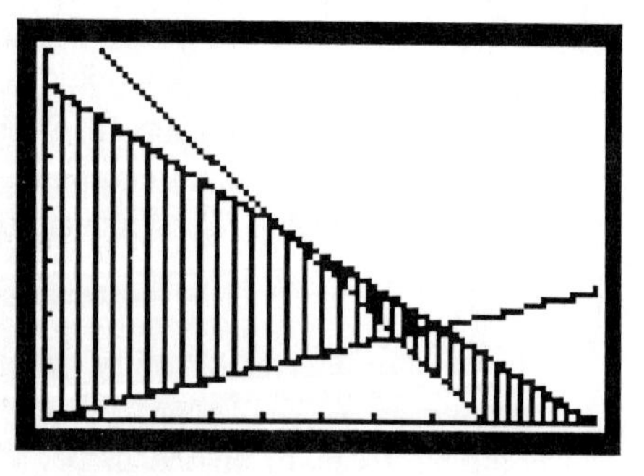

FIGURE 1.145

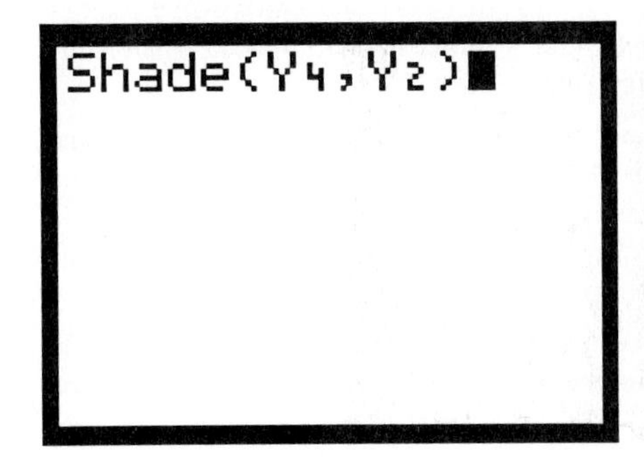

FIGURE 1.146

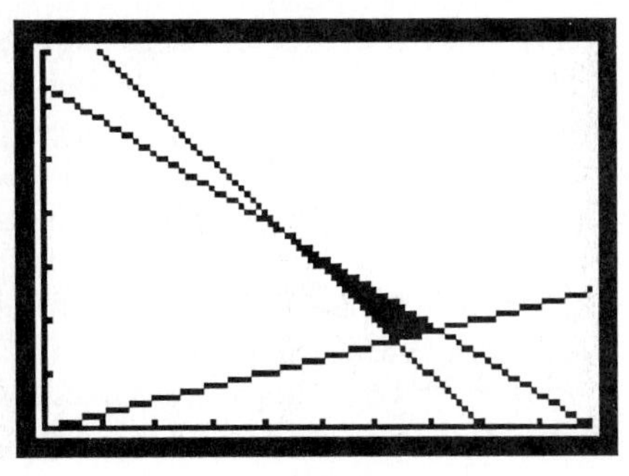

FIGURE 1.147

represented by the inequality $4y \geq x$. Solve for y and enter the companion equation on the Y= menu: $y \geq .25x$; $\mathbf{Y_3 = .25X}$. Figure 1.143 shows the graph of all three lines, with a viewing rectangle of [0, 1000] by [0, 700].

The solution set for this system of three linear inequalities is the same area represented before, but now with a new lower bound, $\mathbf{Y_3 = .25X}$. There are two methods to shade this region. The first method uses two shade commands with different resolutions. Figure 1.144 shows the commands entered on the **home screen** (executed one at a time), and Figure 1.145 shows the resulting shading.

The other method of shading is to create a fourth function specifying conditions for $\mathbf{Y_1}$ and $\mathbf{Y_3}$: $\mathbf{Y_4 = Y_1(Y_1 \geq Y_3) + Y_3(Y_3 \geq Y_1)}$. Figure 1.146 shows the command used to shade the graph, and Fig. 1.147 shows the resulting shading.

Test points in the shaded region by substituting them into all three of the inequality statements at the same time. Do you get three true statements?

Problems

1. Graph the following system of inequalities in two variables using one or two **Shade** commands. State the points you used from inside and outside the shaded regions to check your solution, and show numerically that these points produce true or false statements. Clear the graphing screen between each problem.

   a. $y < 7 - x$ and $y > 2x + 1$
   b. $y > x + 5$ and $y > 3x - 4$
   c. $y < 5 - x$ and $y < 5x - 1$
   d. $y < 2$ and $y > -2$
   e. $y > 6 + 2x$ and $y > 3x - 5$
   f. $y < x + 1$ and $y > x - 1$
   g. $y > 3x - 7$ and $y < 5x + 2$

In problems 2–6, solve by graphing in two variables, and state the point you used from the shaded region to check your solution. Also, give the values that make sense in the problem (i.e. whole numbers, integers, rationals, or reals).

2. Video On Wheels charges a \$10 membership fee and \$3 for each video rented, while Today's Best Video charges a \$20 membership fee and \$2.50 for each video rented. When is the Video On Wheels a better bargain than Today's Best Video?

3. The MTC Phone Company charges \$0.30 for the first minute and \$0.10 for each additional minute or part of a minute, while the Dash Phone Company charges \$ 0.20 for the first minute and \$0.13 for each additional minute or part of a minute. Use a graph to explain when each of the two phone companies is a better buy. Back your argument with an algebraic solution showing when the two companies charge the same amount.

4. If the length of a rectangle is 10 cm more than its width, and its perimeter is less than 100 cm then what are some possible dimensions of the rectangle? What is the maximum length?

5. The Kelley-Shorts stadium will only hold 10,000 people. Tickets for a rock concert are \$3 for students and \$5 for adults. Revenues from ticket sales must exceed \$40,000 before the promoters will make any profit. What is the maximum number of student tickets that can be sold? What is the maximum amount of profit that can be made? Would you expect to make this amount? Explain. How would you alter ticket prices to help the profit margin yet still get lots of people to attend? Explain how this would change the amount of profit expected.

6. Suppose the cost of producing baseballs is \$2.25 and the overhead must be kept under \$1000. If the baseballs are sold for \$2.50, how many would you have to sell to ensure a profit? Could you begin production if you only had \$5000? Explain.

7. The **Shade** command has several features that you should be aware of. You can start and stop the shading at any **X**-value you wish. On the **home screen** enter the command **Shade(Y_1,Y_2,#,X_1,X_2)**, where # is the shading factor (# = 1 is the best), X_1 is the value to start shading, and X_2 is the value to end the shading. Hence the command **Shade(X^2, X+3, 2, 1, 3)** will shade the region between the graphs $Y_1 = X^2$ and $Y_2 = X + 3$ in resolution 2 only from $X = 1$ to $X = 3$.

Graph the following inequalities for the intervals indicated. Experiment with various resolution settings.

a. $x < 2x + 1$ from $x = 0$ to $x = 3$ **b.** $0 < x$ from $x = -2$ to $x = 5$

c. $x > x^2$ from $x = 5$ to $x = 9$ **d.** $0 < 9 - x^2$ from $x = 0$ to $x = 3$

e. $-2 < x < 5$

8. Graph the following groups of three equations and shade the interior region with multiple shade commands. Repeat the problems using the technique outlined in problem 7.

a. $y = 2x + 8$ $y = -3x + 5$ $y = x - 10$

b. $2x - 3y = 4$ $y - x = 3$ $x + y = 10$

9. Shade the following figures using multiple **Shade** commands. Repeat the problems using the technique outlined in problem 7.

a. The triangle with coordinates (-3, 2), (2, 3) and (0, -5)

b. The quadrilateral with coordinates (-2, 3), (3, 4), (4, -3) and (-3, -2)

10. The [TRACE] cursor can be used as a "counter" as well as for finding specific values on a function. Set the [RANGE] to a friendly integer viewing rectangle, and use $Y_1 = 2X$ for even integers and $Y_2 = 2X + 1$ for odd integers to graphically investigate the following. Use the [TRACE] cursor to help complete the following conjectures, and prove them algebraically.

a. The sum of two odd numbers is . . . .

b. The sum of two even numbers is . . . .

c. The sum of an odd number and an even number is . . . .

FOR ADVANCED ALGEBRA STUDENTS

11. Graph the following system of equations in two variables using one **Shade** command, and state the points used to check your solution set.

    a. $y \geq 2x - 8$ and $y > 5x + 2$ b. $y < 5x - 8$ and $y \leq x + 7$

    c. $y < 8 - 3x$ and $y \leq 6x + 1$ d. $y \geq 5 - x$ and $y \geq 3 - 2x$

12. Shade the regions bounded by the following systems of inequalities, and give the points you used to check your solution.

    a. $3x + 2y \leq 5$ $5x + 2y > 6$ $y - x < 10$

    b. $y - x \leq 6$ $x + y \leq 8$ $4y - x + 1 > 0$

    c. $2x + y + 7 \leq 0$ $3y - x \geq 5$ $y > x$

    d. $3x + 2y < 12$ $x - y < 1$ $x < 6$

    e. $x \geq 0$ $y \geq 0$ $2x + 3y < 4$ $x - y < 2$

13. Suppose you are a salesperson in a store that sells a "better" and a "best" quality product. The "better" product sells for $10, and the "best" product sells for $15. To make your monthly bonus you must (1) sell more than 100 items, (2) have more than $1000 in sales, and (3) sell more than twice as many "best" products as "better" products. What combinations of "better" and "best" products can you sell to make your bonus? What is the minimum number of "better" products that you can sell? Write an inequality and graph it to solve the problem; then check it algebraically and numerically.

14. A coin collector has been collecting "wheat" pennies and "buffalo head" nickels for years. The total number of coins is less than 100, and their face value is greater than $3. Also, the number of pennies is less than 10 larger than 3 times the number of nickels. What are some possible combinations of the two coins? What is the maximum possible number of nickels? Write an inequality and graph it to solve the problem; then check it algebraically and numerically.

15. The Three Ring Circus is in town and charges $1.50 for children and $4.50 for adults. The main tent only holds 1000 people but every three children must be accompanied by an adult (i.e. the ratio of children to adults cannot exceed 3 to 1). What combination of tickets could be sold so that revenue exceeds $2000? Write an inequality and graph it to solve the following problems; then check your results algebraically and numerically.

    a. What is the maximum number of children's tickets that could be sold?

    b. What is the maximum amount of profit from one performance?

16. Graph the following system of equations in two variables and state the points used to check your solution set.

    a. $x^2 - x + 1 < x + 3$ b. $x^2 - x + 1 \geq x + 3$

    c. $x^2 + 2x - 5 \leq 3 - x^2$ d. $\sin x \geq x - 1$

e. $\sqrt{x} > x$ **f.** $\sin x < x$

g. $\sin x \leq \cos x$

17. If the length of a rectangle is 20 in. more than its width, then when is the perimeter greater than its area? What if the length is three times the width?

18. The following are examples of supply and demand curves:

supply: $y = .72x - .1$, demand: $y = 10 + .2x - .01x^2$.

They usually are determined by the fact that as prices (y) go up, the sellers increase production (x) (more supply) and the consumer buys less (less demand). At what point is there an equilibrium (i.e. when does supply equal demand)? When is the supply greater than the demand? When is demand greater than supply? Explain the best business strategy if you are

a. the manufacturer; **b.** the consumer.

19. Suppose you want to drive 500 miles in less than 9 hours, including a 30-minute lunch stop. What speeds could you average for the trip if you cannot exceed 75 miles per hour? Write an inequality and graph it to solve the problem; then check it algebraically and numerically.

20. In two variables, graphically investigate the following inequalities for different values of a, and make conclusions about when these statements are true.

a. $|x + a| \leq |x| + |a|$ **b.** $|x - a| \geq |x| - |a|$

2

Quadratic Equations and Functions

2.1 Exploring Quadratic Equations

An expression of the form

$$a_nx^n + a_{n-1}x^{n-1} + a_{n-2}x^{n-2} + \cdots + a_2x^2 + a_1x + a_0$$

is called a **polynomial in *x***, where *n* stands for a nonnegative integer and the terms a_n, a_{n-1}, and a_{n-2} stand for real numbers, called **coefficients** of the different terms. If the coefficient of the first term, a_n, does not equal zero, then *n* is called the **degree** of the polynomial, since it is the highest power of the variable.

A **quadratic equation** is a polynomial equation with degree 2. For example,

$$y = 2x^2 - 3x - 1 \quad \text{and} \quad y^2 + 3y = x$$

are both quadratic equations. Equations with degree 1 are called linear equations. The equation $y = 2x - 5$ is a linear equation. Quadratic equations are also called **second degree** polynomials since their degree is 2. The equations $x^2 + 3y^2 = 9$, $y = \frac{2x-3}{x^2+5}$, $y = \sin^2 x + \cos^2 x$, and $y = \sqrt{x^2 - 4}$ are not quadratic equations. Why?

EXPLORATION 1: A **quadratic function** is a quadratic equation whose graph is a set of points where no two points have the same first coordinate. For example, use the

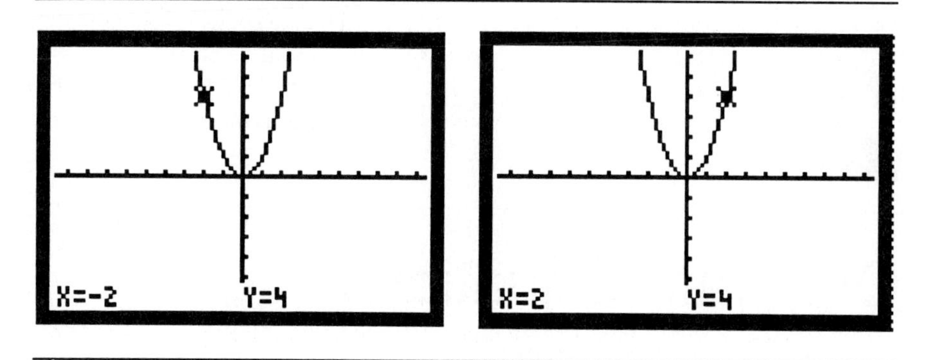

FIGURE 2.1

FIGURE 2.2

TI-81 to draw the graph of the quadratic function $y = x^2$ in the viewing rectangle [-9.6, 9.4] by [-6.4, 6.2]. Activate the TRACE cursor and investigate the points that make up the graph. Figure 2.1 shows the graph with the TRACE cursor on the point (-2, 4). Are there any other points with the same first (x) coordinate of -2? Are there other points on the graph with the same second (y) coordinate of 4?

Figure 2.2 shows another point with the same second coordinate, (2, 4). In order for two points to have the same first coordinate, the two points would lie on a vertical line. Can you pass a vertical line through the graph of $y = x^2$ at any point and touch the graph at two or more points? Try it! The answer is no. Figure 2.3 shows several vertical lines through the graph of $y = x^2$.

Use the **Line** command (interactively) from the DRAW menu to draw vertical lines. Notice that when you draw a vertical line only the y-coordinate changes. The x-coordinates of all points on a vertical line are the same. None of these vertical lines touches the graph at more than one point. This procedure is called the **vertical line test**. In general, if you draw a vertical line through the graph of an equation at any

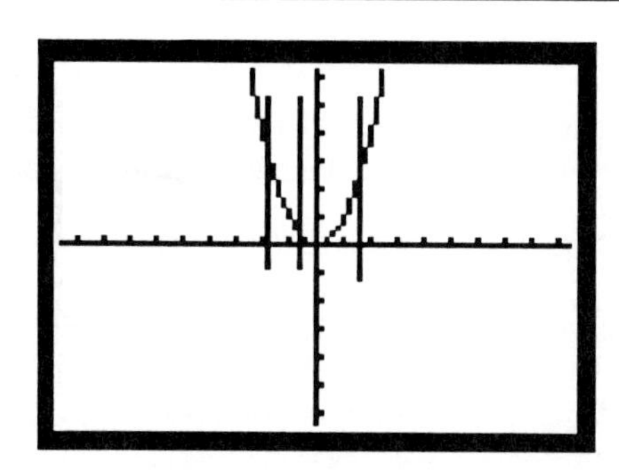

FIGURE 2.3

point and the line touches no more than one point of the graph at any time, then the equation represented by the graph is a function. Conversely, if a vertical line drawn through the graph of an equation touches two or more points on the graph, then the equation represents a **relationship**, not a function. A function is more restrictive than a simple relationship. ◊

EXPLORATION 2: Draw the graph of the quadratic equation $y^2 = x$. Does this quadratic equation represent a quadratic function?

When we entered linear functions on the [Y=] menu, we had to solve the equations for y in terms of x. The same is true for quadratic equations, as well as all other types of equations. Solving for y yields

$$y^2 = x$$
$$y = \pm\sqrt{x},$$

or

$$y = \sqrt{x} \quad \text{and} \quad y = -\sqrt{x}.$$

The single quadratic equation becomes two equations since the product of two positive or two negative values is positive. To graph the original equation, enter the two resulting equations as

$$\mathbf{Y_1 = \sqrt{X}} \quad \text{and} \quad \mathbf{Y_2 = -\sqrt{X}}$$

First draw the graph in **Sequence** mode (see Fig. 2.4).

Notice that the first piece of the graph draws completely before the second piece starts. The graphs are drawn in sequence as they are listed on the [Y=] menu. Set the TI-81 in **Simul** mode and press [GRAPH] again. This time the graphs are drawn simultaneously. Does this graph represent a quadratic function? How can you tell? Move the [TRACE] cursor to the point (4, 2) on the upper branch of the graph. Press the [▼] arrow to shift from $\mathbf{Y_1}$ to $\mathbf{Y_2}$. What are the coordinates of the point on

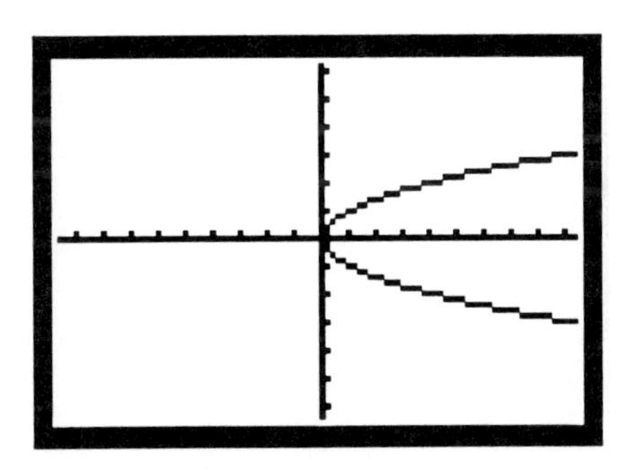

FIGURE 2.4

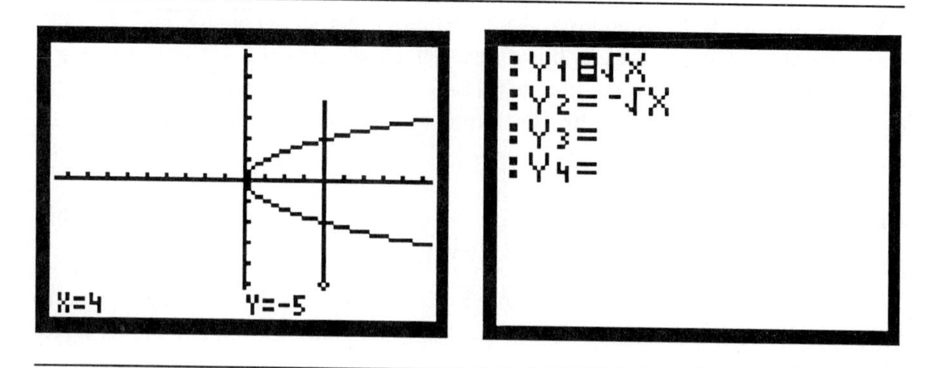

FIGURE 2.5

FIGURE 2.6

Y_2? What is true about the points (4, 2) and (4, -2)? Could you pass a vertical line through both these points at the same time? Figure 2.5 shows a vertical line drawn through these points with the **Line** command.

What would happen if we graphed just one branch of the graph at a time? Activate the Y= menu, and turn off $\mathbf{Y_2}$ by placing the cursor on the equal sign (=) and pressing ENTER. You can identify which functions on the Y= menu are active by looking for the equal signs in **reverse video** (darkened). Figure 2.6 shows the Y= menu with $\mathbf{Y_1}$ active and $\mathbf{Y_2}$ inactive, and Fig. 2.7 shows the resulting graph.

When you press GRAPH, the calculator will ignore $\mathbf{Y_2}$ and only graph $\mathbf{Y_1}$. To activate $\mathbf{Y_2}$ again, repeat the procedure. The equal sign acts like a **toggle** switch to turn the function on or off.

If we consider only the upper part, is this the graph of a function? Does it pass the vertical line test? If we consider only the lower branch, is this the graph of a function? Draw the graph of the lower branch by turning $\mathbf{Y_1}$ off and $\mathbf{Y_2}$ on. The graph

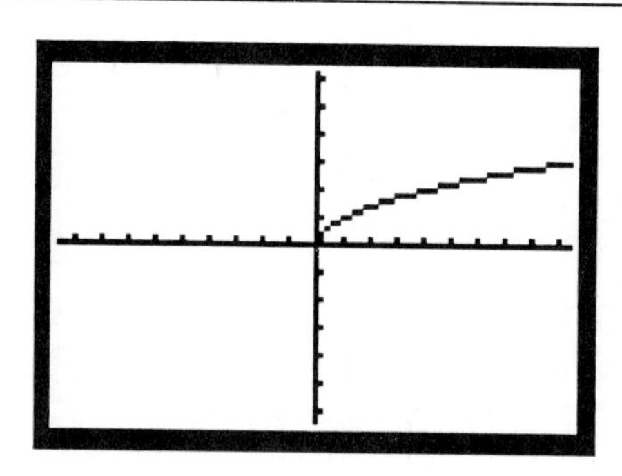

FIGURE 2.7

of the lower branch, like the upper branch, represents a function. But when we consider the complete graph of the equation $y^2 = x$, we have a quadratic relationship, not a quadratic function in x. One reason we do not have a function is that the graph of the equation does not pass the vertical line test. Another clue to this result is the fact that we need to enter the equation in two parts in order to draw the complete graph of the equation. We will examine other relationships later in this manual. For the remainder of this chapter we will deal with quadratic functions. ◊

EXPLORATION 3: Draw the graph of the quadratic function $y = 2x^2 + 3x - 2$.

(a) Investigate the ***x*-intercepts** of the graph (points where the graph crosses the x-axis).

Figure 2.8 shows the graph of $\mathbf{Y_1 = 2X^2 + 3X - 2}$ in the basic friendly viewing rectangle [-4.8, 4.7] by [-3.2, 3.1]. The TRACE cursor is positioned on one of the x-intercepts. Figure 2.9 shows the x-intercept. What do you notice about the y-coordinate of these two points? Why would the y-coordinate have to be zero at the x-intercept? All the points on the x-axis have a y-coordinate of zero!

(b) How many intercepts are there?

Are there any other x-intercepts for this function? Activate the TRACE cursor and trace to the right a long way by holding the ▶ arrow key down. Even after the cursor leaves the screen, the values shown at the bottom of the screen are correct. Eventually the screen will **scroll** to the left as the x values get larger than **Xmax**. If you TRACE to the right far enough, the graph will scroll completely off the screen. Figure 2.10 shows the graph scrolled to the left edge of the screen.

The y values shown on the screen are the values you would get if you substituted the x value shown into the original function. For example, the point (6.9, 113.92) is shown in Fig. 2.10. Substitute this point into the original equation and the result is a true statement:

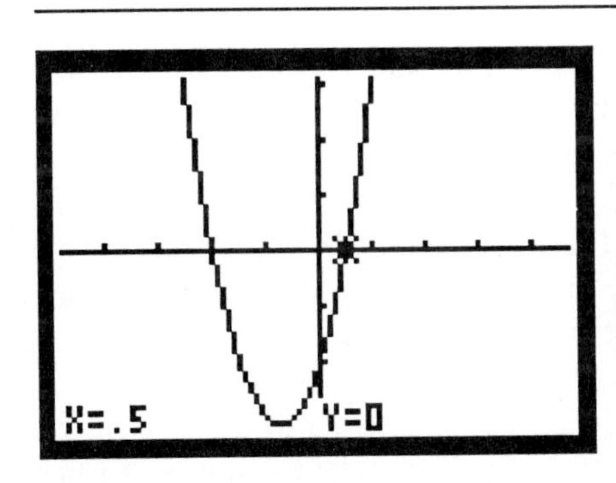

FIGURE 2.8

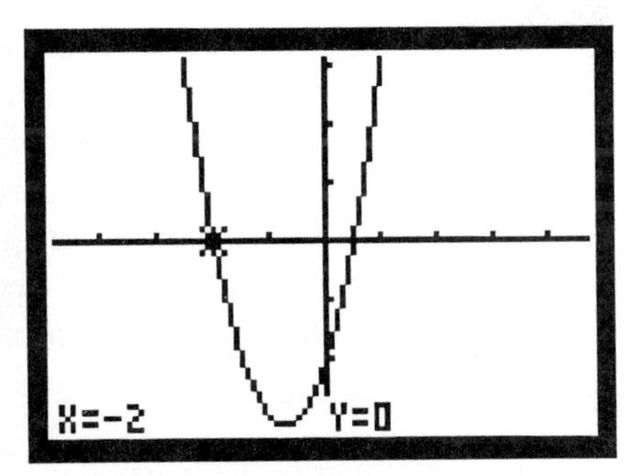

FIGURE 2.9

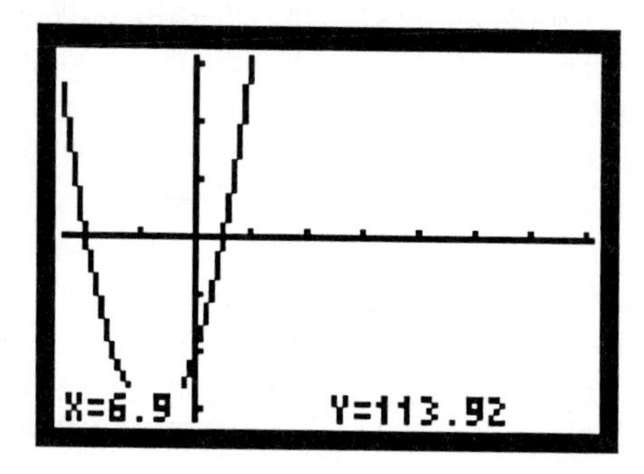

FIGURE 2.10

$$
\begin{aligned}
y &= 2x^2 + 3x - 2 \\
113.92 &= 2(6.9)^2 + 3\,(6.9) - 2 \\
113.92 &= 2\,(47.61) + 20.7 - 2 \\
113.92 &= 95.22 + 20.7 - 2 \\
113.92 &= 113.92 \qquad \textbf{True} \text{ statement}
\end{aligned}
$$

In a mathematical sense, we say that for a given x value, the y values represent the function value at x, or **f(*x*)**(read "*f* of *x*"), to the limits of the decimal display of the calculator. When using the TRACE cursor, the values shown on the screen are x and the corresponding $f(x)$ for the given function. This is not true for the **screen cursor.**

What happens to the y value as x gets larger and larger? In order for this graph to have another x-intercept, the y values would have to approach zero again. Does this ever happen? Let the x values get as large as you wish, say 100 or 200, by moving the TRACE cursor to the right. What happens to the y values?

Move the TRACE cursor to the left as far as you wish, say to $x = -100$ or -200. What happens to the y values? Can you make a general conclusion about the **end behavior** of the graph as the x values get very large or very small (movement to the left on the number line means numbers are getting smaller)? The y values keep getting larger and larger. This means that there are no more than two x-intercepts for this quadratic function.

To confirm this conjecture visually, draw the graph of $y = 2x^2 + 3x - 2$ in a much larger viewing rectangle, say [-100, 100] by [-1000, 10,000], and use the TRACE cursor to check values (see Fig. 2.11). Explore other larger viewing rectangles to confirm that the graph never crosses the x-axis again.

(c) What is the meaning of these points?

The x-intercepts of the function $y = 2x^2 + 3x - 2$ represent the values of x that make $y = 0$. Alternately, the x-intercepts represent the values of x that make the

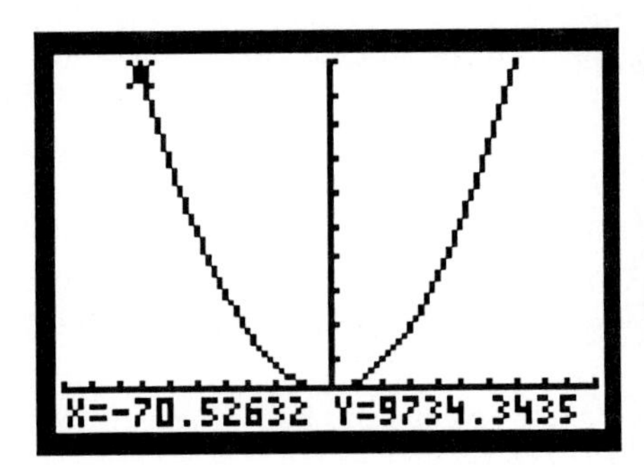

FIGURE 2.11

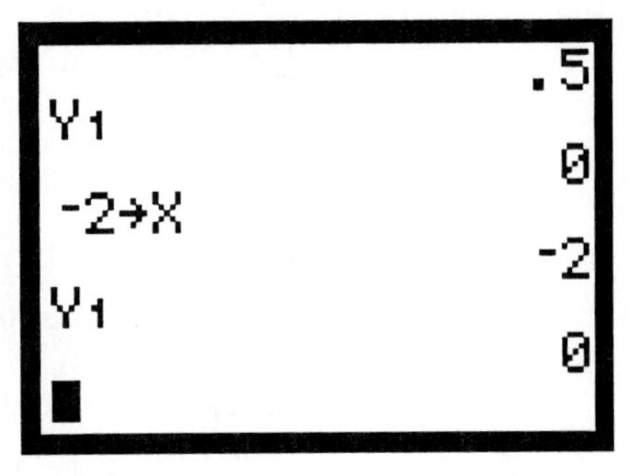

FIGURE 2.12

equation $2x^2 + 3x - 2 = 0$ true. These are also known as the **zeros, or roots, of the function.** For this function, the values $x = .5$ and $x = -2$ are the solutions, zeros, or roots, of the equation $2x^2 + 3x - 2 = 0$:

$$\begin{aligned} 2(.5)^2 + 3(.5) - 2 &= 0 \\ 2(.25) + 1.5 - 2 &= 0 \\ .5 + 1.5 - 2 &= 0 \\ 0 &= 0 \quad \textbf{True} \end{aligned} \qquad \begin{aligned} 2(-2)^2 + 3(-2) - 2 &= 0 \\ 2(4) - 6 - 2 &= 0 \\ 8 - 6 - 2 &= 0 \\ 0 &= 0 \quad \textbf{True} \end{aligned}$$

With the function $\mathbf{Y_1 = 2X^2 + 3X - 2}$ stored on the Y= menu, use the STO ▶ key to evaluate the function at X = .5 and X = -2. Figure 2.12 shows the results. When solving problems modeled by quadratic or other functions, the zeros or roots often represent the solutions.

(d) How do these points relate to the algebraic representation of this function?

Another way to find the zeros of a quadratic equation is to find the prime factors and set each of them equal to zero. The prime factors of $2x^2 + 3x - 2 = 0$ are found by several methods outlined in your Algebra textbook. We will not review these methods here. In factored form,

$$2x^2 + 3x - 2 = (2x - 1)(x + 2).$$

To check this, multiply the right side of the equation and collect like terms:

$$(2x - 1)(x + 2) = 2x^2 + 4x - x - 2 = 2x^2 + 3x - 2.$$

To find the roots, or zeros, of the equation $2x^2 + 3x - 2 = 0$, examine the factored form

$$(2x - 1)(x + 2) = 0.$$

By the **zero factor property**, either $(2x - 1) = 0$, or $(x + 2) = 0$, since their product equals zero. Solve the two equations:

$$2x - 1 = 0 \qquad\qquad x + 2 = 0$$
$$2x = 1 \qquad\qquad x = -2$$
$$x = .5$$

These are the same solutions we got from the graph. When we factor a quadratic equation equal to zero and solve for values of x that make a true statement, we are finding the values of x where the graph of the quadratic function crosses the x-axis. The x-intercepts of the graph are the visual representation of the roots, or zeros, of the function.

By factoring, we know

$$2x^2 + 3x - 2 = (2x - 1)(x + 2).$$

We could use either of these representations to draw the graph of the original function. Figure 2.13 shows the [Y=] menu with the two different versions entered.

You can select each equation separately to show that they produce the same graph. Or, you can graph both functions at the same time and then use the [TRACE] cursor to show that every point on both graphs coincides. With both graphs drawn and the [TRACE] cursor active, when you press the [▼] arrow, the [TRACE] cursor jumps from one graph to the next. If the x- and y-coordinates printed on the screen do not change, then the individual points are on both of the graphs. This procedure works for all the points on the graphs. Figure 2.14 shows the [TRACE] cursor output when the [▼] arrow is held down with both graphs drawn.

In many cases, especially in applications problems, the roots of a quadratic equation are not easy to read from the graph, and the equation does not factor. For these cases, other solution techniques are used. ◊

APPLICATION EXPLORATION: If an object is thrown upward vertically, its height at any time after the start of the motion is given by the quadratic function

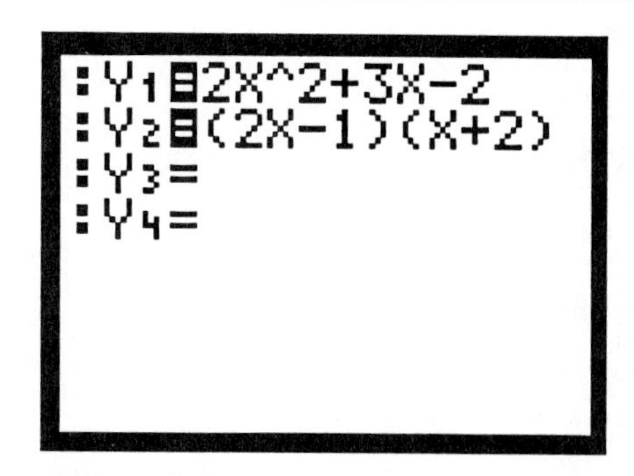

FIGURE 2.13

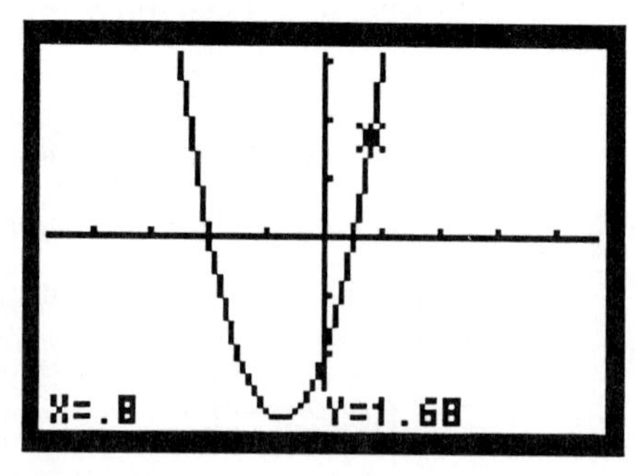

FIGURE 2.14

$$H(t) = -4.9t^2 + v_0t + s_0,$$

where H = the height, -4.9 = the acceleration due to gravity in meters per second, t = time, v_0 = the initial velocity, and s_0 = the starting height of the object. This is a function, so we say that ***height is a function of time*** because the height of the object depends on how much time has passed since it was thrown.

Your science class is investigating the effects of gravity on falling bodies. As an experiment, the teacher stands on the edge of the roof of the school and propels a baseball vertically upward at initial velocities from 15 to 30 meters per second (in increments of 5 meters per second). The path of the ball is such that it will just miss the edge of the roof on the way down. By measuring, the class determines that the roof of the school is 15 meters high. Figure 2.15 shows the problem situation.

Each member of the class uses a stopwatch to time the interval from when the ball is thrown to when it strikes the ground. The experiment is repeated several times and data are recorded for each trial. Table 2.1 gives the values from one student's notebook.

Draw the graph of this experiment for each of the four initial velocities. Using the mathematical formula, calculate the the time it **should** take, with error at most .01, for the baseball to hit the ground at each of the four initial velocities. Compare these results to the experimental results shown in Table 2.1. Explain any differences you find.

To draw these graphs on the TI-81, we must enter the functions in terms of the variables **X** and **Y**. Other variables may be used in the function definition; however, they represent the one value stored in that memory register. The value of **X** changes

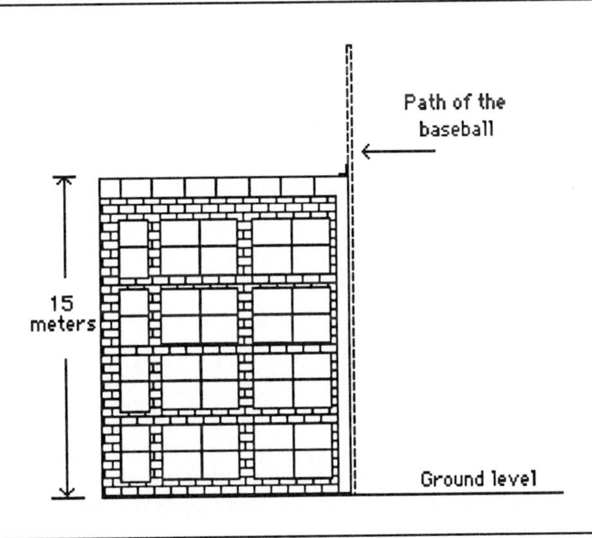

FIGURE 2.15

TABLE 2.1
TIME UNTIL THE BASEBALL HITS THE GROUND (SECONDS)

	Initial velocity (meters per second)			
Experimental trials	*15*	*20*	*25*	*30*
1	4.0			
2	3.8			
3	4.1			
4		4.8		
5		4.7		
6		4.9		
7			5.5	
8			5.6	
9			5.8	
10				6.5
11				6.8
12				6.7
Theoretical results				

from **Xmin** to **Xmax** as the function is graphed. For this problem situation, let **X** represent time, t, and let **Y** represent height, H, from the original equation. For an initial velocity of 15 meters per second and a starting height of 15 meters, the first equation would be

$$\mathbf{Y_1 = -4.9X^2 + 15X + 15.}$$

Figure 2.16 shows the [Y=] menu with all four equations and Fig. 2.17 shows the resulting graphs in the viewing rectangle [0, 9.5] by [-10, 75].

Remember, these graphs do not represent the flight path of the ball, but the height of the ball as a function of time. The x-axis represents ground level (why?). The point at which the graph crosses the x-axis represents the time when the baseball hits the ground. The ball hits the ground when the height is zero. The equations we need to solve are

$$0 = -4.9x^2 + 15x + 15,$$
$$0 = -4.9x^2 + 20x + 15,$$
$$0 = -4.9x^2 + 25x + 15,$$
$$0 = -4.9x^2 + 30x + 15.$$

Zoom Box Method

The first solution method we will use is like the zoom in technique we used in the first chapter, except we will use the power of the TI-81 to do the zooming.

Select **1: Box** from the [ZOOM] menu. A **screen cursor** appears at the center of

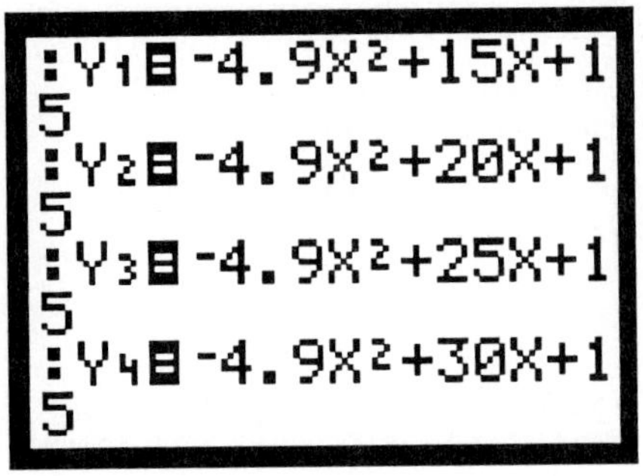

:Y1=-4.9X²+15X+15
:Y2=-4.9X²+20X+15
:Y3=-4.9X²+25X+15
:Y4=-4.9X²+30X+15

FIGURE 2.16

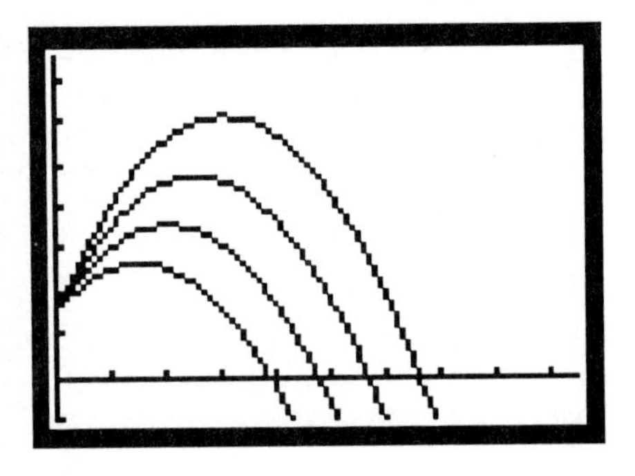

FIGURE 2.17

the graphics screen. The cursor represents one corner of a rectangular box. We want to place the box so that the point of interest is inside the box when it is drawn. First we will zoom in on the solution of $0 = -4.9x^2 + 15x + 15$ by placing the box around the point on the graph representing the solution. Figure 2.18 shows the first corner placed above and to the left of the x-intercept of the graph. Once the first corner is in position, press ENTER to set the corner. (This is like nailing this corner to the graph.)

When you press ENTER the cursor changes to a small box. Stretch this box horizontally and vertically with the arrow keys until it captures the point of interest. Figure 2.19 shows the box in position. When the box is in position, press ENTER and the small box will become the entire graphics screen. After zoom in, the new viewing rectangle will be given on the RANGE menu. This process may be repeated as many times as necessary to find a value to a given degree of accuracy.

- **Notes on using the Box command:** The first corner of the zoom box may be set anywhere in relation to the point of interest. Use the arrow keys to stretch the box

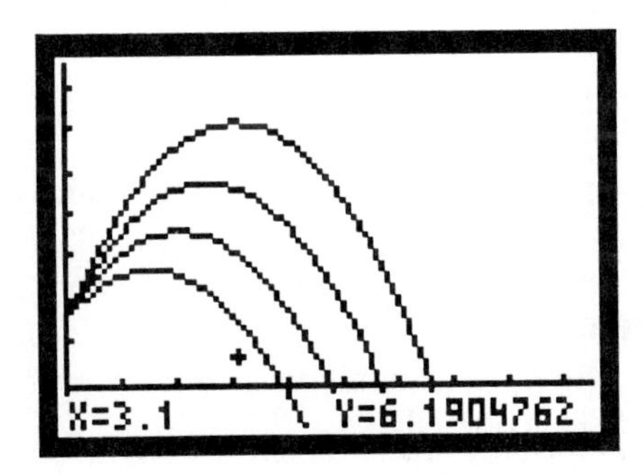

FIGURE 2.18

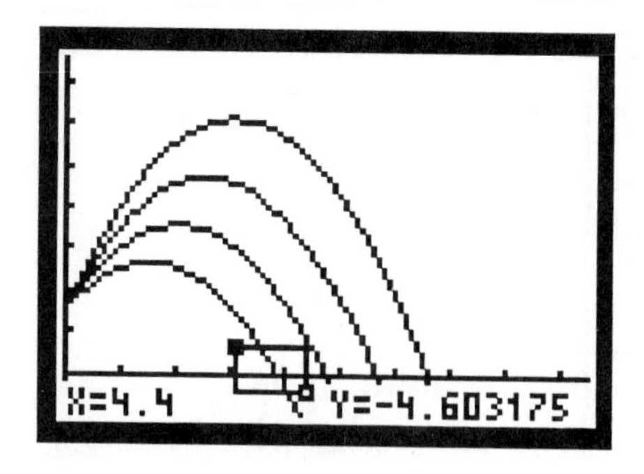

FIGURE 2.19

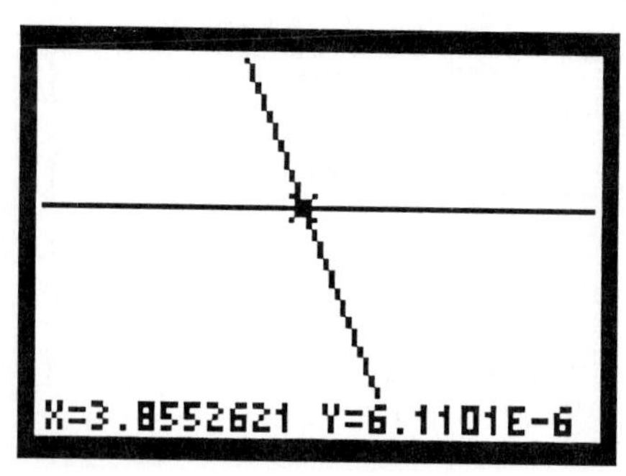

FIGURE 2.20

in the appropriate directions. At any time before the second corner is set, the zoom box may be erased by pressing [GRAPH]; this allows mistakes to be corrected before zooming. If you zoom in "too far," a **Range Error** will occur. If this happens, the [RANGE] must be changed from the [RANGE] menu before another graph can be drawn. If you are zooming in on only one function, you can speed up the graphing process by deactivating other functions on the [Y=] menu.

To solve the equation to an error of less than .01, we need to zoom in until the distance across the screen in both the x and y direction is less than .01. Repeat the **Box** command several times. Check your progress by referring to the [RANGE] menu. Since each user's zoom box may be slightly different, Fig. 2.20 shows an example of a viewing rectangle within the limits of error of .01. Figure 2.21 shows the [RANGE] menu for this viewing rectangle.

To check the size of the viewing rectangle, compute **Xmax – Xmin** and **Ymax – Ymin** (see Fig. 2.22). These [RANGE] variables are found on the [VARS] menu under the **RNG** heading.

Activate the [TRACE] cursor, as in Fig. 2.19, to estimate the root at the time **X = 3.8552621** seconds when the height above the ground is **Y= 6.1101E – 6.** (*Note:* This **Y** value is given in scientific notation: 6.1101E – 6 = 6.1101×10^{-6} = 0.0000061101; given an accuracy of .01, this value represents 0.00.) Given an error of at most .01, we can report that the time the baseball, with an initial velocity of 15 meters per second, hits the ground should be between 3.85 and 3.86 seconds. This compares very well with the experimental results given in Table 2.1. Causes for differences between the experimental and theoretical results could be timing errors, air resistance, initial velocity slightly different from 15 meters per second, or the starting height not measured correctly.

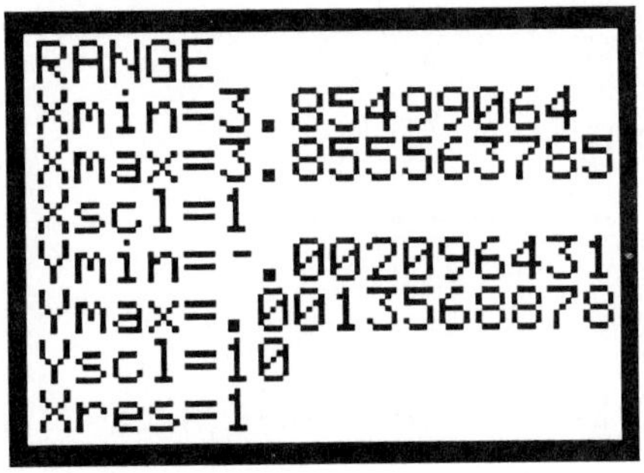

```
RANGE
Xmin=3.85499064
Xmax=3.855563785
Xscl=1
Ymin=-.002096431
Ymax=.0013568878
Yscl=10
Xres=1
```

FIGURE 2.21

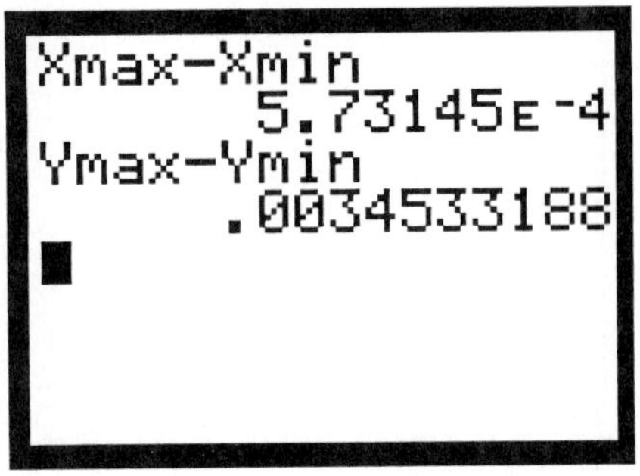

```
Xmax-Xmin
        5.73145E-4
Ymax-Ymin
        .003453318
```

FIGURE 2.22

Zoom to Point Method

The TI-81 has another automatic zoom in technique represented by the **2: Zoom In** command from the [ZOOM] menu. This technique could be more accurately called "zoom to point," since the user must identify a point to use for the center of the zoomed in viewing rectangle.

To use this command, first select **4: Set Factors** from the [ZOOM] menu. This option allows the user to set the zoom in and zoom out factors for both the horizontal and vertical direction. If you set **XFact = 10** and **YFact = 10**, then when you zoom in or out, the screen will be changed by a factor of 10 in the horizontal and vertical directions. Setting factors that are not equal means that you will zoom in or out in one direction more than in the other. This is called **differential zooming** and can be used effectively in certain circumstances.

Next, select **2: Zoom In** from the [ZOOM] menu. A **screen cursor** will appear on the graphics screen. Place this cursor **on** the point of interest and press [ENTER]. The viewing rectangle will be changed by the factors you set, and the new viewing rectangle will be centered (approximately) at the point you selected. Figure 2.23 shows the **2: Zoom In** cursor ready to zoom on the x-intercept of $y = -4.9x^2 + 15x + 15$. The other equations have been turned off, not erased from the [Y=] menu. This **screen cursor** can also be used to estimate the coordinates of the point of interest. The point (3.9, -1.029) is a good estimate of the time for the ball to hit the ground, given the size of the viewing rectangle of [0, 9.5] by [-10, 75].

After the graph is drawn in the new viewing rectangle, the **2: Zoom In** cursor appears at the center of the graphics screen. To zoom in again, simply place the cursor on the point of interest and press [ENTER] . This process can be repeated as many times as you wish. Check your progress by referring to the [RANGE] menu. Figure 2.24 shows the [TRACE] cursor on the screen after zooming in several times. The results are similar to the **2: Zoom Box** technique: **X = 3.855265**, and **Y = -5.928E – 5**.

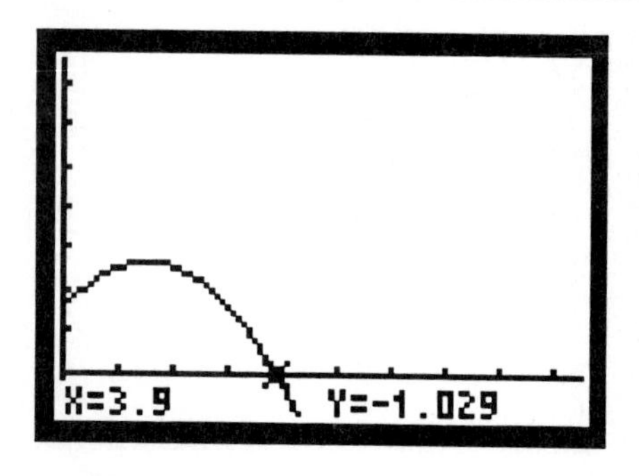

FIGURE 2.23

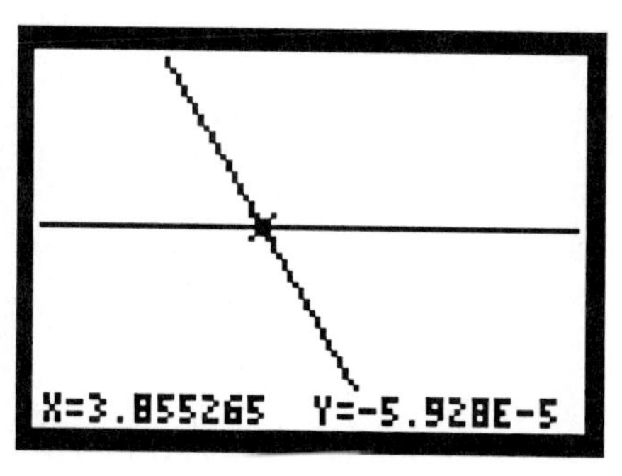

FIGURE 2.24

- **Caution!** If you try to set the zoom factors (**4: Set Factors**) too large, you may zoom past the point of interest and lose it from the graphics screen. This is like using a microscope set on too high a power to find a detail on the slide. Zoom factors in the range from 4 to 10 are the most effective while still providing a "safe" margin for error.

Quadratic Formula Method

If you complete the square of the general quadratic expression

$$0 = ax^2 + bx + c$$

and solve for x, the result is the **quadratic formula:**

$$x = \frac{-b \pm \sqrt{b^2 - 4ac}}{2a},$$

where a, b, and c are the coefficients of the terms of the quadratic expression, and $a \neq 0$. Refer to your algebra textbook for development of the technique called **completing the square** and using it to develop the quadratic formula.

For the equation $0 = -4.9x^2 + 15x + 15$, $a = -4.9$, $b = 15$, and $c = 15$. Applying the formula

$$x = \frac{-15 \pm \sqrt{15^2 - 4(-4.9)(15)}}{2(-4.9)}.$$

The "±" symbol indicates that there are really two values represented by this equation. This is consistent with the fact that there can be two x-intercepts for a quadratic equation. Figure 2.25 shows this expression entered on the screen of the TI-81. The second phrase was produced by using the **Replay** function (▲ arrow) and changing the "+" to a "–" (subtraction).

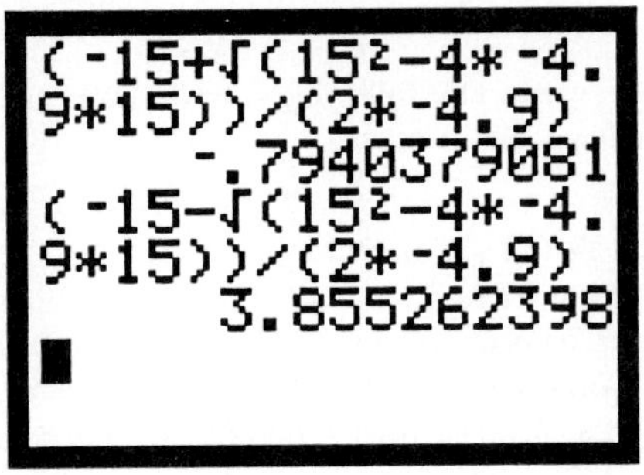

```
(-15+√(15²-4*-4.
9*15))/(2*-4.9)
       -.7940379081
(-15-√(15²-4*-4.
9*15))/(2*-4.9)
        3.855262398
```

FIGURE 2.25

The value $x = 3.855262398$ is consistent with the graphical solutions; compare this with the value **X = 3.8552645** from one of the zoom in procedures. The other value $x = -.7940379081$ is mathematically correct; however, this value represents **time** in the problem setting, and a negative time has no meaning. Why didn't we see this other x-intercept on the graph? Could you?

Solve the other three equations by a variety of the three methods. The completed table will look like Table 2.2. ◊

TABLE 2.2
TIME UNTIL THE BASEBALL HITS THE GROUND (SECONDS)

	Initial velocity (meters per second)			
Experimental trials	*15*	*20*	*25*	*30*
1	4.0			
2	3.8			
3	4.1			
4		4.8		
5		4.7		
6		4.9		
7			5.5	
8			5.6	
9			5.8	
10				6.5
11				6.8
12				6.7
Theoretical results	3.86	4.73	5.64	6.59

Problems

1. In the interactive mode, connect the following points in order and use the vertical line test to see if it would be a function.

   a. (1, 2), (3, 5), (-1, 2) b. (-2, -3), (1, 1), (5, 3)

2. Graph $\mathbf{Y_1 = X - 1}$, and $\mathbf{Y_2 = X + 2}$, and $\mathbf{Y_3 = (X - 1)(X + 2)}$ in a friendly window with an aspect ratio of 1. Using the TRACE cursor, move from one line to the other, and then to the quadratic function. Compare the y values for any value of x to convince yourself that the product of the two lines is the quadratic. That is, $\mathbf{Y_1 * Y_2 = Y_3}$. Pay particular attention to the signs of the values. What happens when they are both negative; both positive; one positive and one negative? What if they are both small; both large; one small and one large? What happens when $x = 1$ or $x = -2$? What if we try $(x + 2)$ and $(3 - x)$? Try other pairs of lines, and form a quadratic function to investigate the effect of graphing their product. Make a generalization about this process.

3. Using ZOOM **1: Box,** solve the following quadratic equations graphically with an error of at most .01. Check your results algebraically.

   a. $x^2 - 5x + 6 = 0$ b. $6x^2 - 13x + 6 = 0$

   c. $3x^2 - 4x - 2 = 0$ d. $x^2 + 2x + 3 = 0$

4. Using ZOOM **2: Zoom In,** solve the following quadratic equations graphically with an error of at most .01. Check your results algebraically.

   a. $x^2 - 5x + 6 = 0$ b. $6x^2 + 7x - 5 = 0$

   c. $x^2 + 9x + 7 = 0$ d. $6 - 8x - x^2 = 0$

5. Instead of factoring to find the rational roots of an equation, you can test the roots by entering the equation on the Y= menu, using the STO▶ key to store possible roots in **X**, and then evaluating $\mathbf{Y_1}$ on the home screen (see Fig. 2.12). You can find the possible roots in the same way you find possible factors. What value must you get for $\mathbf{Y_1}$ to have a root? Solve the following quadratic equations in this manner.

   a. $x^2 + 5x + 6 = 0$ b. $6x^2 - 7x - 5 = 0$

   c. $3x^2 + 2x - 8 = 0$ d. $4x^2 - 15x + 9 = 0$

6. a. Given that the roots are -2 and 3, write the quadratic function and check by graphing. Is this the only quadratic function with roots of -2 and 3? Write at least two more quadratic functions with the same roots. Can you make a general rule?

   b. Find the quadratic equation with roots of -2 and 3 and a y-intercept of 4. Can you write more than one quadratic equation with these specifications? Why or why not? Explain.

7. For the roots of a general quadratic equation $ax^2 + bx + c = 0$, the sum of the roots is given by $\frac{-b}{a}$, and the product of the roots is given by $\frac{c}{a}$. Check the formulas by finding the roots of the following equations graphically.

   a. $-x^2 + 3x - 2 = 0$ b. $3x^2 - 4x - 4 = 0$

   c. $x^2 + 4x - 1 = 0$ d. $-2x^2 - 4x + 5 = 0$

8. Graph several quadratics ($y = ax^2 + bx + c$) in which $a = b$ and both are four times as large as c (i.e. $y = 4cx^2 + 4cx + c$), and make a conjecture about the number of roots. Prove your conjecture algebraically.

9. When an object is thrown straight up into the air, the height is found by the function $H(t) = -16t^2 + v_0t + s_0$, where t represents the time, v_0 is the initial velocity, and s_0 is the starting height of the object. If an object is thrown from ground level at a velocity of 65 feet per second, find the following:

a. Write and graph the equation.

b. What values make sense in the problem situation?

c. When will the ball hit the ground?

d. What if the ball was thrown from a building 30 feet high? Redo (a)–(c) for this new situation.

10. The perimeter of a rectangle is 500 inches.

a. Using x to represent the length, express both width and area in terms of x.

b. What is the area of the rectangle in terms of x?

c. Graph the equation for the area in an appropriate [RANGE].

d. If the area is 500 square inches, what are the dimensions of the rectangle?

e. If the length is 100, what is the area of the rectangle?

11. The sum of the first x terms of the number pattern $10 + 8 + 6 + \cdots$ is given by the function $S(x) = 11x - x^2$, where x represents the number of terms. How many terms does it take to get a sum of zero? Confirm your answer graphically, algebraically, and numerically.

Write an equation for the problem situations in problems 12–15. Solve algebraically and support your answer graphically and numerically.

12. Find two numbers if their difference is 8 and the sum of their squares is 100.

13. If you leave school at 3 P.M. walking north at 2 miles an hour, and your friend leaves at the same time walking east at 3 miles per hour, when are you 1 mile apart?

14. If the length of a rectangle is 7 more than its width, and the area is 200, then what are the dimensions of the rectangle?

15. A swimming pool with dimensions 20 by 30 feet is surrounded by a sidewalk of uniform width, x. Find the possible widths for the sidewalk if the total area of the sidewalk is to be greater than 200 but less than 360 square feet.

16. For each of the following algebraic solutions examine the solution; show by substitution that the values found are solutions to the original equations. Then solve by graphing both sides of the original equations. Do you get the same results? Explain any differences you see. Could you solve the original equations algebraically in another way?

a.
$$x^2 + 2x + 1 = x + 1$$
$$(x + 1)(x + 1) = x + 1$$
$$x + 1 = 1$$
$$x = 0$$

b.
$$x^2 - 2x = 0$$
$$x^2 = 2x$$
$$x = 2$$

17. Graph $y = x^2$ in a friendly window, and use the **Line** command to draw a line from (0, -1) to any point (a, a^2) on the graph. The area of the triangle formed by the line drawn, the x-axis, and the vertical line $x = a$ is given by $\frac{a^5 - a}{2(a^2 - 1)}$. Check this by finding the length of the height and base of the triangle with the **screen cursor** and using the formula for the area of a triangle, $A = \frac{1}{2}bh$.

18. Graph $y = x^2$ in the integer window [0, 95] by [0, 6300]. Find any numbers whose square ends in that same number. For example, $5^2 = 25$. You can TRACE as high as you wish.

19. Graph $y = x^2$ in a [-9.4, 9.6] by [0, 63] window. Use the **Line** command to draw a line connecting any two points on the parabola. Make sure the y values correspond to the x values on the graph of $y = x^2$. The absolute value of the y-intercept of this line will be the same as the product of the two x-coordinates of the endpoints of the line. Try this for several different pairs of points. Explain why this is true algebraically.

20. The total surface area of a right circular cylinder is $A = 2\pi r^2 + 2\pi rh$. Find different radii of the cylinder if the height is 5 inches. What radius would produce a surface area of 25π square inches?

FOR ADVANCED ALGEBRA STUDENTS

21. In theory, prime numbers occur randomly and cannot be generated by a formula. The following formulas seem to return only prime numbers for integer values of x. Use the friendly integer window [0, 95] by [0, 63] and the TRACE cursor to investigate these functions for many integer values of x.

I. Do these formulas return only prime numbers for integer values of x?

II. Do these formulas give a complete list of the prime numbers?

III. Explain algebraically why these formulas must fail in both (a) and (b).

a. $P(x) = x^2 - x + 17$ **b.** $P(x) = x^2 - x + 41$

22. Given the quadratic equation $ax^2 + bx + c = 0$, where a, b, and c are real numbers, then:

- If $b^2 - 4ac < 0$, then the equation has two **imaginary roots;**
- If $b^2 - 4ac = 0$, then the equation has one **real root;**
- If $b^2 - 4ac > 0$, then the equation has two **real roots.**

The quantity $b^2 - 4ac$ is called the **discriminant** of the quadratic equation. The value of the discriminant can be tested by entering the phrase $\mathbf{Y_1 = B^2 - 4AC}$ on the Y= menu, using the STO ▶ key to enter the appropriate values for **A, B,** and **C,** and evaluating $\mathbf{Y_1}$ on the **home screen.** Figure 2.26 shows the Y= menu and Fig. 2.27 shows the **home screen** with the discriminant evaluated for **A** = 2, **B** = -3, and **C** = 5. Repeat this procedure and classify the following quadratic function by the types of roots they have. Confirm your analysis by graphing the equations in the general form $\mathbf{Y_2 = AX^2 + BX + C}$, as seen in Fig. 2.26.

a. $y = x^2 - 5x - 7$ **b.** $y = 3x^2 - 4x + 2$ **c.** $y = x^2 + 8x + 16$

23. Solve the following systems of equations within .01 graphically, and check algebraically.

a. $y = x^2 - 5x + 1$ and $y = 5x - 2$

b. $y = 5x^2 + 3x - 6$ and $y = 1 - x^2$

c. $y = x^2 + 3x - 4$ and $y = 2(x^2 - 1)$

24. Solve the following inequalities with error at most .01. Use the **Shade** command to show the solution set graphically (i.e. to show when one graph is above/below the other). Check your results algebraically.

a. $x^2 - 4x + 3 < x + 3$

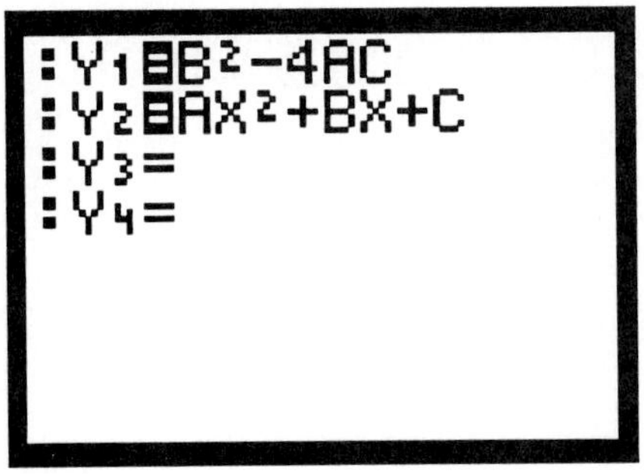

:Y1=B²−4AC
:Y2=AX²+BX+C
:Y3=
:Y4=

FIGURE 2.26

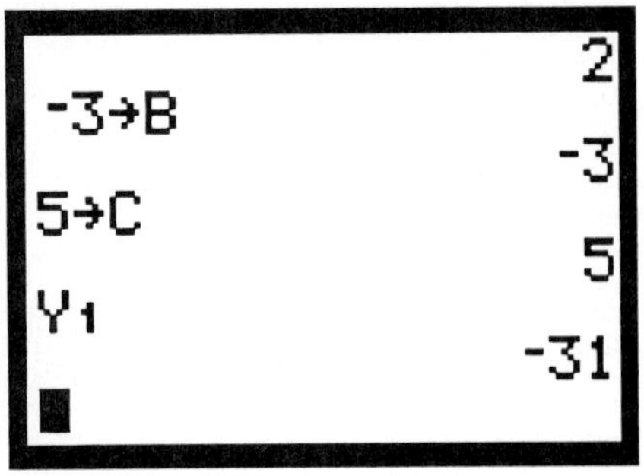

-3→B
-3
5→C
5
Y1
-31

FIGURE 2.27

b. $x^2 - 4x + 3 \le -x^2 + 4x - 3$

c. $2x^2 + 3x + 4 \ge 3x - 2$

25. Find two consecutive integers such that four times the first is equal to the square of the second. Using **8: INT** from the [ZOOM] menu or a friendly integer window will assist in reading the [TRACE] cursor.

26. For the following equations, find when the supply is equal to the demand:

Supply $= 5 - x^2$ and Demand $= x^2 - 10$.

When is the supply greater than the demand? Shade the area between the two curves.

27. Profit is equal to the revenue minus the cost, or $P = R - C$. Given the following equations for revenue and cost, determine the break-even point by graphing a single equation:

$R = 5000x - 20x^2$ and $C = 300{,}000 - 1500x$.

Next solve the same problem by graphing an inequality in two variables. Did you get the same answer?

28. Determine a, b, c so the graph of $y = ax^2 + bx + c$ contains the points (1, 8), (-1, 2), and (3, 6). (*Hint*: Solve a system of three linear equations in three unknowns.) Check your

FIGURE 2.28

DATA
x1=1
y1=8
x2=-1
y2=2
x3=3
y3↓6

FIGURE 2.29

FIGURE 2.30

solution graphically by entering the points in the **statistical memories** and plotting the points using the **2: Scatter** command from the STAT **Draw** submenu. To enter points in the **statistical memories,** first select the **2: ClrStat** command from the **Data** submenu of the STAT menu (see Fig. 2.28). This command clears all previous statistical values from the memory. Next select the **1: Edit** command from the same submenu to enter the coordinates of the points. Figure 2.29 shows the three points entered in the statistical memories.

Set an appropriate viewing window and select the **2: Scatter** command to plot the three points on the screen (see Fig. 2.30). Once you find values for a, b, and c, use the **DrawF** command to draw the parabola that will hit all three points. (*Note*: If you graph the equation from the Y= menu, the points will be erased before the graph is drawn.)

29. Graph the following quadratic function involving absolute values, and make a conjecture about the shape of the graphs in general. How does the number of real roots affect the shape of these graphs? Explain.

a. $y = |x^2 - 3x - 6|$ **b.** $y = |x^2 + 3x + 1|$ **c.** $y = |9 - 5x - 3x^2|$

30. There are other "quadratic" type equations that we want to explore. Graphically solve the following within .01, and check your solutions algebraically. (*Note*: $\sin^2 x = (\sin x)^2$; use the form **(sin X)^2** on the TI-81.)

a. $x^4 - 5x^2 + 6 = 0$ **b.** $\sin^2 x + 2 \sin x + 1 = 0$

2.2 Exploring Families of Quadratic Functions

The simplest form of a quadratic function is $y = x^2$. In mathematical terms, this is known as the **canonical form** of the quadratic function. By experimenting with the coefficients of the quadratic function we will create an entire **family of functions.** The function $y = x^2$ is called the **parent function,** since all other forms come from this basic form. There are many families of functions. The quadratic family is one of the most important.

EXPLORATION 1: Draw the graph of $y = x^2$, and investigate the effects of changing the coefficient of the x^2 term. Organize your observations and make a generalization that will help you predict the basic shape of the graph of a quadratic function.

Enter the function on the [Y=] menu: $\mathbf{Y_1 = X^2}$. Draw the graph in the viewing rectangle [-9.6, 9.4] by [-6.4, 6.2] (i.e. factor of 2 for the **Range Program**). Figure 2.31 shows the graph. For this series of examples, place the TI-81 in **Sequence** mode to draw multiple graphs one at a time.

Use the [TRACE] cursor to investigate the numerical values represented by the graph. Notice that each point on the graph has a **symmetric** point on the other side of the y-axis, which has the same y-coordinate and the opposite x-coordinate (see Figs. 2.31 and 2.32). Is there any point that does not have a symmetric companion point? Trace along the graph to the point (0, 0). This one point has no companion point. A graph with these symmetric properties is called a **parabola**. The one point that has no symmetric companion point is called the **vertex** of the parabola. If we draw a vertical line through the parabola at the vertex, we create the **line of symmetry** of the parabola. In the case of $y = x^2$, the line of symmetry is the y-axis; the equation of this line is $x = 0$. This is not always true for other parabolas.

In the quadratic function $y = x^2$, the coefficient of x^2 is 1. In general we represent the coefficient of the x^2 term as a: $y = ax^2$. Let's experiment with different values for this coefficient. Add the graph of $y = 2x^2$ to the screen ($\mathbf{Y_2 = 2X^2}$). Figure 2.33 shows the two graphs.

What do you notice about the new graph? Is it symmetric in the same way as the first graph? What is the vertex of the new graph? What is the equation of the line of symmetry? Use the [TRACE] cursor to investigate the new graph. When the [TRACE] cursor is first activated, it appears on the first active function, $\mathbf{Y_1}$. To shift to other active functions, press the [▲] or [▼] arrows. Place the [TRACE] cursor on the point (1.6, 2.56) on the graph of $\mathbf{Y_1 = X^2}$ (see Fig. 2.34). Is this correct? (Yes, $1.6^2 = 2.56$.) For the same **X** value, what will the value of $\mathbf{Y_2 = 2X^2}$ be? Press the [▼] arrow and the [TRACE] cursor jumps to the other graph at the point (1.6, 5.12):

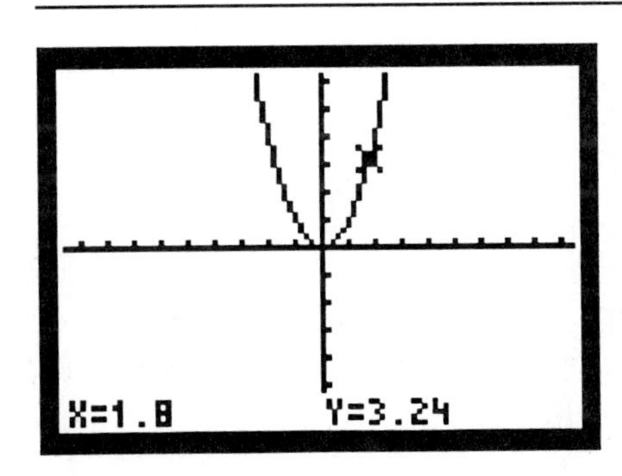

FIGURE 2.31

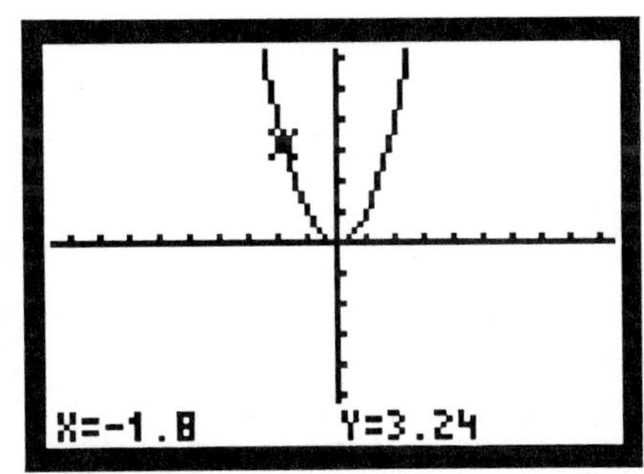

FIGURE 2.32

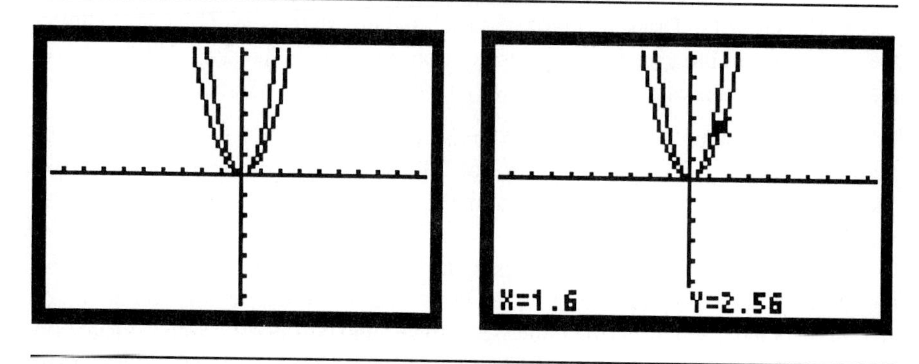

FIGURE 2.33 FIGURE 2.34

$2(1.6)^2 = 5.12$. Figure 2.35 shows this view. The y-coordinate of this point represents twice the value of the basic function, since $y = 2x^2$.

Investigate this situation for other values of **X**. For $\mathbf{X = 2, X^2 = 4}$ and $\mathbf{2X^2 = 8}$; for $\mathbf{X = -3, X^2 = 9}$ and $\mathbf{2X^2 = 18}$. In each case the y-coordinate on the graph of $y = 2x^2$ is double the y-coordinate on the graph of $y = x^2$. Does this work for all points? Try many values for **X** both on and off the screen. Try the point (0, 0). Does the same relationship hold true?

Add another graph to the screen: $\mathbf{Y_3 = 4X^2}$. Before you press the GRAPH key, predict how the new graph will look based on your previous experience. Were you correct? Figure 2.36 shows the three graphs. Activate the TRACE cursor, and predict the y-coordinates for the graphs of $\mathbf{Y_2 = 2X^2}$ and $\mathbf{Y_3 = 4X^2}$ when $\mathbf{X = -4}$. ($\mathbf{Y_2 = 32}$ and $\mathbf{Y_3 = 64}$; see Figs. 2.36 and 2.37.) Explain.

Do these three graphs have any points in common? What are the vertex and the equation of the line of symmetry for the parabola $y = 4x^2$? What is the relationship

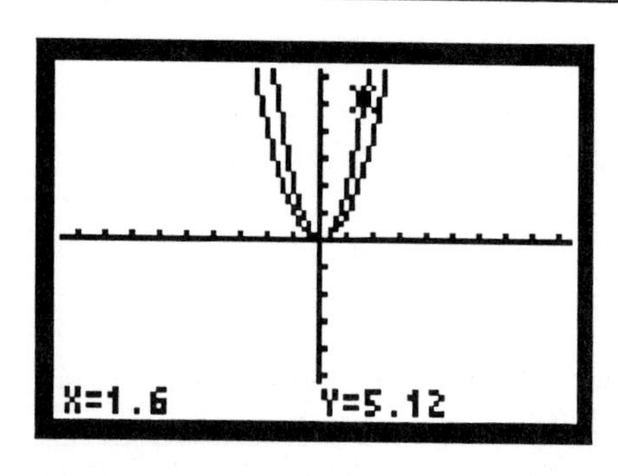

FIGURE 2.35

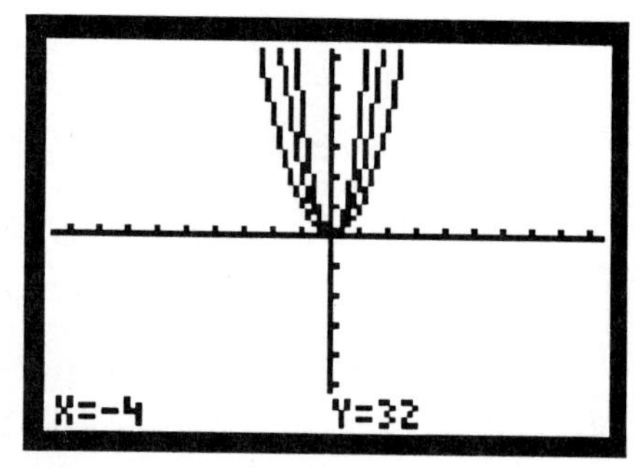

FIGURE 2.36

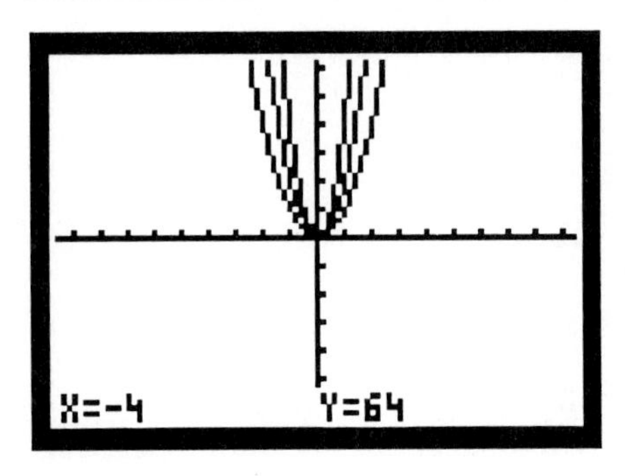

FIGURE 2.37

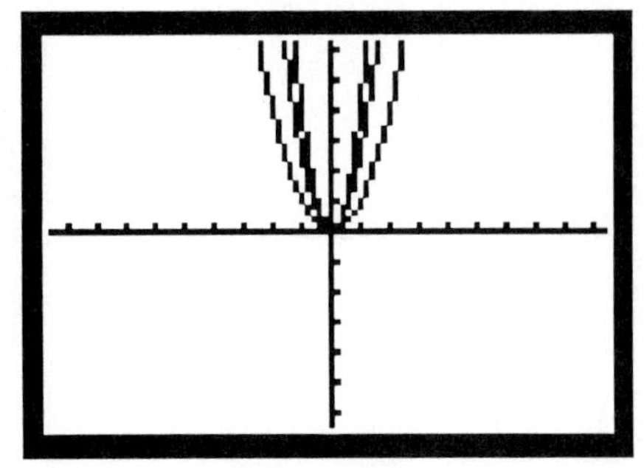

FIGURE 2.38

between the y-coordinates of the points on the graph of $y = x^2$ and $y = 4x^2$? Explain. Predict how the graphs of $y = 10x^2$, $y = 20x^2$, and $y = 100x^2$ will appear. Graph them on the TI-81 and see if you are correct.

Predict the shape of the graph of $y = 2.5x^2$. Try it; were you correct? Figure 2.38 shows the graphs of $y = x^2$, $y = 2.5x^2$, and $y = 4x^2$ on the same screen. Notice that the graph of $y = 2.5x^2$ lies between the other two graphs in the same way that 2.5 lies between 1 and 4 on the number line.

We say that a **vertical stretch** is applied to the parent graph of $y = x^2$ to become the graph of $y = 2x^2$ or $y = 4x^2$. Can you make a general rule for the shape of the graph of $y = ax^2$ for $a \geq 1$? Does this vertical stretch apply to every point on the graphs? Why does the point with an x-coordinate of zero not get stretched like other points?

In general, for values of $a \geq 1$ the graph of $y = ax^2$ is a vertical stretch of "a times" the graph of $y = x^2$. Every point of the graph is changed except points having an x-coordinate of zero. ◊

EXPLORATION 2: Compare the graph of $y = x^2$ with the graphs of $y = ax^2$ for values of a in the range $0 < a < 1$. Make a general rule about the shape of the graph for these coefficients.

Draw the graph of $y = x^2$ and $y = .5x^2$ on the same graphics screen (see Fig. 2.39). What do you notice about the shape of the graph of $y = .5x^2$? What are the vertex and the line of symmetry for this graph? Activate the TRACE cursor and move to the point (-2, 4) on the graph of $y = x^2$. Predict the y-coordinate of the point on the graph of $y = .5x^2$ with x-coordinate of -2. Switch the TRACE cursor to the other graph. Were you correct? See Fig. 2.40.

Notice that the points on the graph of $y = .5x^2$ have y-coordinates that are one-half the value of the y-coordinates on the graph of $y = x^2$. Why? Are there any points that do not follow this pattern? Test many pairs of points to confirm this pattern. Test the point (0, 0); explain what happened.

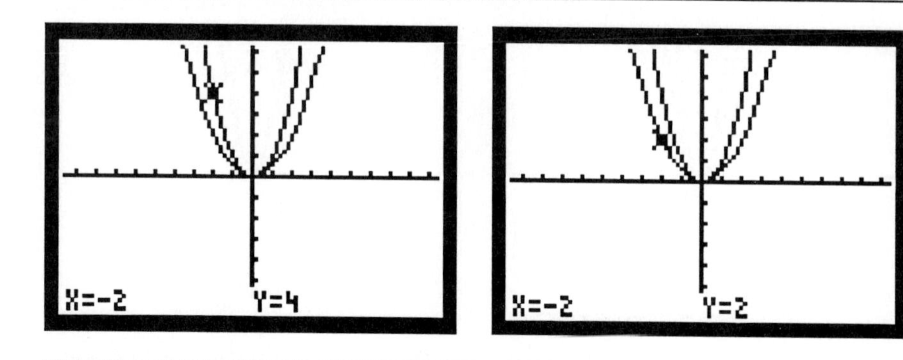

FIGURE 2.39

FIGURE 2.40

Add the graph of $y = .25x^2$ to the graphics screen. Before you draw the graph, predict the shape and location of this new graph. Will it be above or below the graph of $y = .5x^2$? Why? Add the graph of $y = .1x^2$; predict the position of this graph. Figure 2.41 shows all four graphs in the viewing rectangle [-9.6, 9.4] by [-6.4, 6.2].

Move the TRACE cursor to the point (2, 4) on the graph of $y = x^2$. Predict the y value for each point on the other three graphs with an x-coordinate of 2. Figure 2.42 shows the TRACE cursor on the graph of $y = .1x^2$ at the point (2, .4). Were your predictions correct? Try other starting points on the graph of $y = x^2$, and predict the values on the other graphs.

We say that a **vertical compression** is applied to the parent graph of $y = x^2$ to become the graph of $y = .5x^2$, $y = .25x^2$, or $y = .1x^2$. Can you make a general rule or the shape of the graph of $y = ax^2$ for $0 < a < 1$? Does this vertical compression apply to every point on the graphs? Does the point with an x-coordinate of zero not get compressed like other points? What would happen if $a = 0$? Investigate and explain! ◊

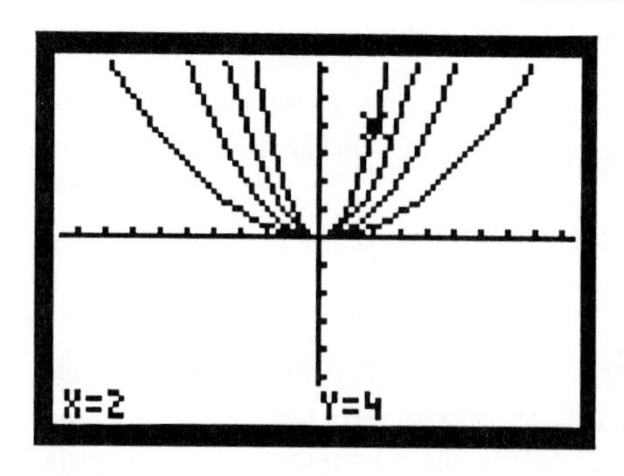

FIGURE 2.41

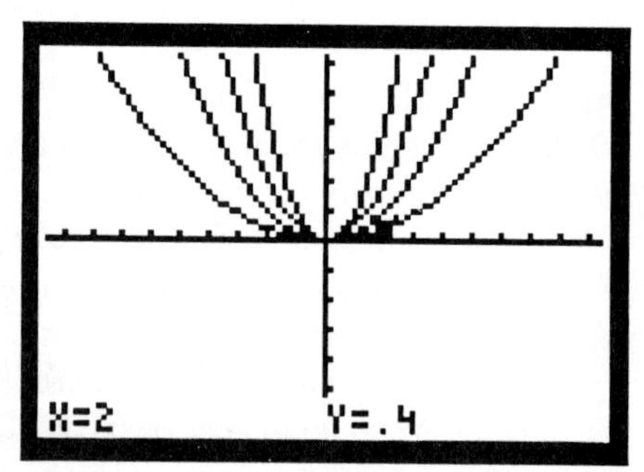

FIGURE 2.42

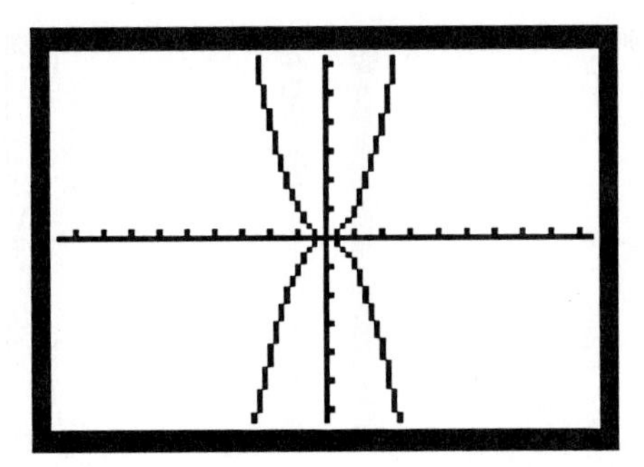

FIGURE 2.43

EXPLORATION 3: Investigate negative values for a in the general quadratic equation $y = ax^2$ by drawing the graphs of $y = x^2$, $y = 2x^2$, $y = -x^2$, and $y = -2x^2$. Investigate negative values for the range $-1 < a < 0$ by drawing the graphs of $y = .5x^2$, $y = -.5x^2$, $y = .1x^2$, and $y = -.1x^2$.

Before any graphs are drawn, make a prediction about the position of the graph of $y = -x^2$. This means that for each x value, the corresponding y value will be the opposite of the x value squared. These y values are just the opposite of the values for the graph of $y = x^2$. We might expect the graph of $y = -x^2$ to be the "opposite" of the graph of $y = x^2$. Figure 2.43 shows these two graphs.

The graph of $y = -x^2$ appears to be the graph of $y = x^2$ **reflected across** the x-axis. Was your prediction correct? What is the vertex of the graph of $y = -x^2$? What is the equation of the line of symmetry? Is the shape of this graph the same as the shape of the graph of $y = x^2$? Does this relationship work for the graphs of $y = 2x^2$ and $y = -2x^2$? Figure 2.44 shows these two graphs. Activate the TRACE cursor on the graph of

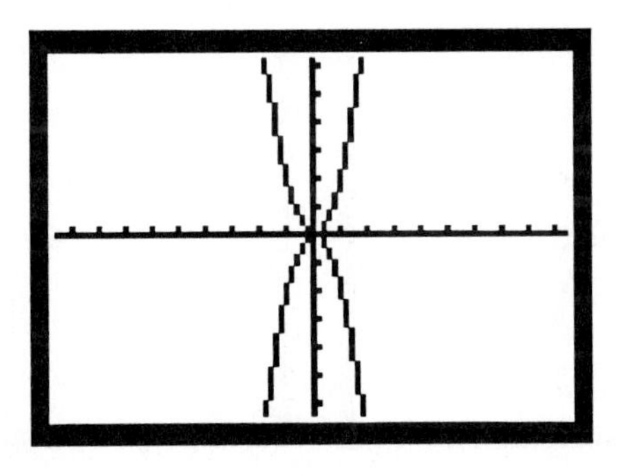

FIGURE 2.44

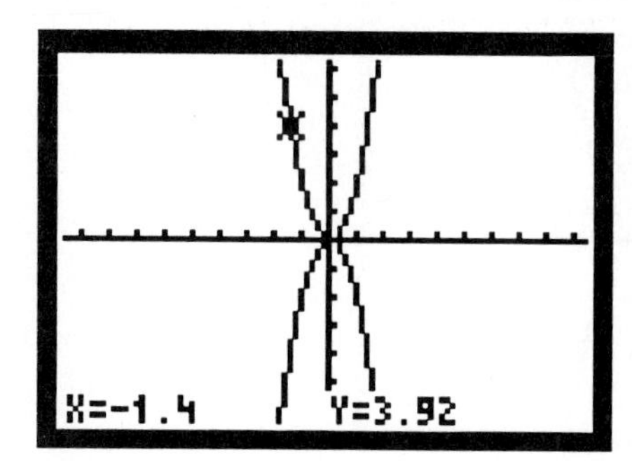

FIGURE 2.45

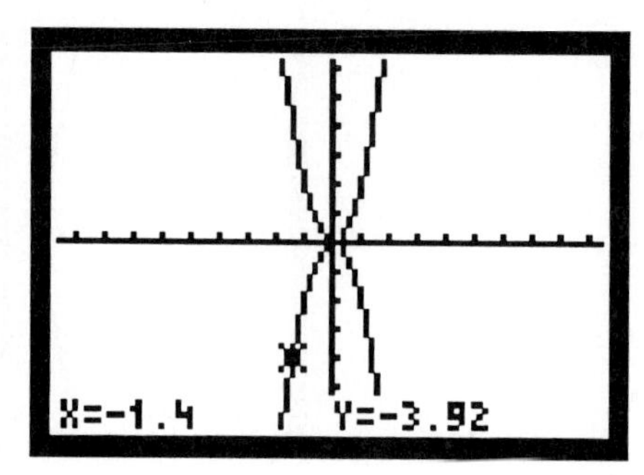

FIGURE 2.46

$y = 2x^2$, and switch to the graph of $y = -2x^2$. Notice that for each point on the graph of $y = 2x^2$, the corresponding point on the graph of $y = -2x^2$ has the opposite y-coordinate. Figures 2.45 and 2.46 show this relationship.

For negative fractional values of a, the same relationship holds true. Figure 2.47 shows the graphs of $y = .5x^2$ and $y = -.5x^2$, and Fig. 2.48 shows the graphs of $y = .1x^2$ and $y = -.1x^2$.

In each case, the graph of $y = ax^2$ with a negative value for a is the same as the graph of $y = ax^2$ with the corresponding positive value for a reflected across the x-axis. The TRACE cursor confirms this hypothesis numerically. When a is positive, the graph opens upward, and when a is negative, the graph opens downward. ◊

EXPLORATION 4: Draw the graphs of $y = x^2$ and $y = x^2 + 2$ on the same graphics screen. Describe how the parent graph is changed into the new graph.

Draw the two graphs on the graphics screen. Figure 2.49 shows these graphs in the viewing rectangle [-9.6, 9.4] by [-6.4, 6.2]. Activate the TRACE cursor and trace to

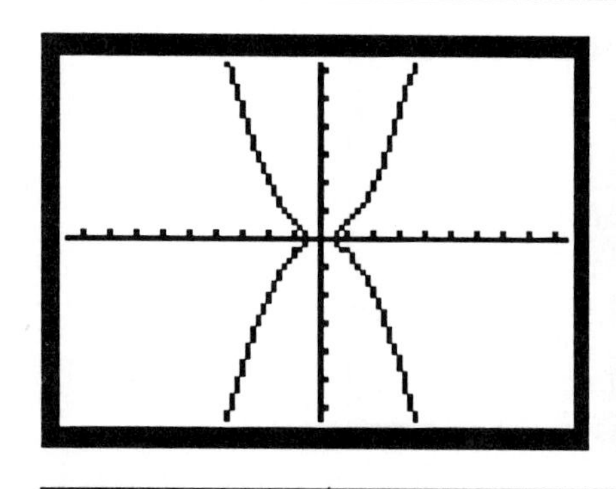

FIGURE 2.47

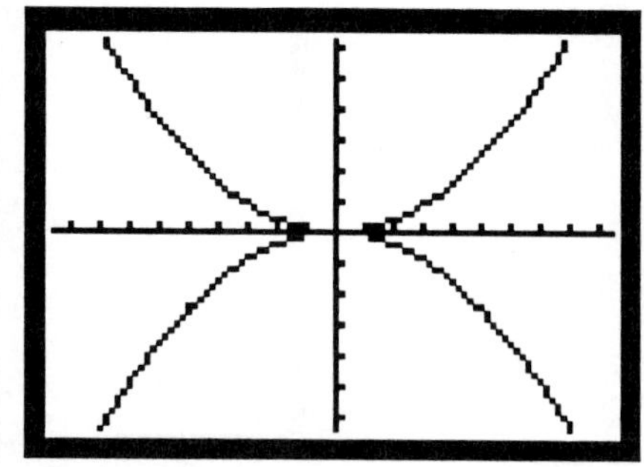

FIGURE 2.48

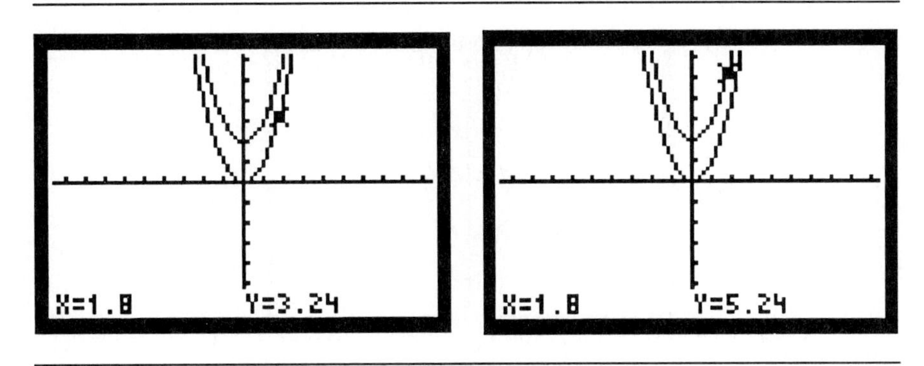

FIGURE 2.49 FIGURE 2.50

the point (1.8, 3.24) on the graph of $y = x^2$. Predict the corresponding point on the graph of $y = x^2 + 2$. Figure 2.50 shows this point. Can you explain this?

Try other pairs of corresponding points on the two graphs. What point corresponds to (-1.2, 1.44) on the graph of $y = x^2$? [Answer: (-1.2, 3.44).] Try the vertex of $y = x^2$, (0, 0). The companion point on the graph of $y = x^2 + 2$ is (0, 2). Can you explain this? See Figs. 2.51 and 2.52.

When we investigated changing the value of a in $y = ax^2$, the point with x-coordinate of 0 did not move. This time the vertex point (0, 0) did move to the point (0, 2). Explain. It seems that each and every point on the graph of $y = x^2$ is shifted 2 units in the vertical direction to get the graph of $y = x^2 + 2$. Does this equation itself tell you this?

Add the graphs of $y = x^2 - 3$ and $y = x^2 - 5$ to the graphics screen (see Fig. 2.53). For a given point on the graph of $y = x^2$, predict the corresponding points on the other graphs. For example, trace to the point (2, 4) on the graph of $y = x^2$. The corresponding points on the other graphs are

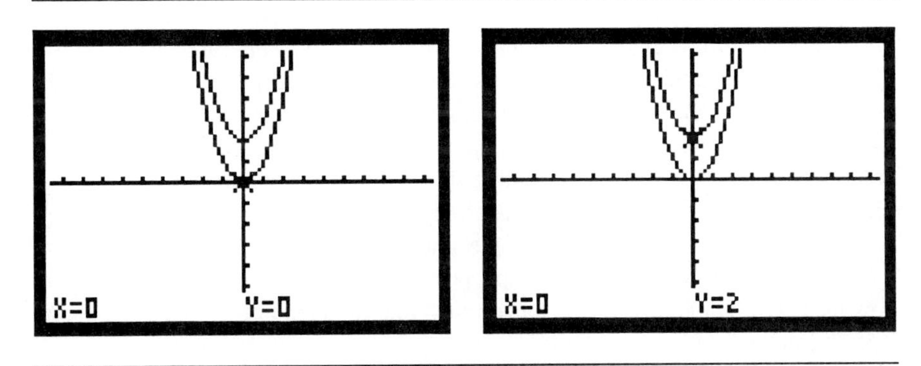

FIGURE 2.51 FIGURE 2.52

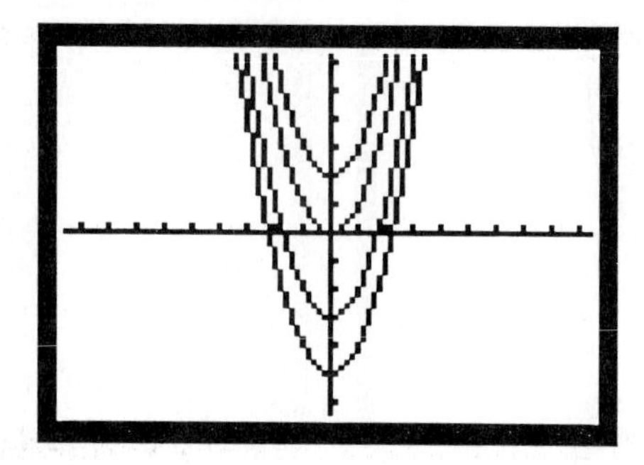

FIGURE 2.53

$y = x^2$	(2, 4)	$(4 + 0 = 4)$
$y = x^2 + 2$	(2, 6)	$(4 + 2 = 6)$
$y = x^2 - 3$	(2, 1)	$(4 - 3 = 1)$
$y = x^2 - 5$	(2, -1)	$(4 - 5 = -1)$

Notice the pattern. In each case the amount added to or subtracted from the y-coordinate of the point on the parent function is the same amount added to or subtracted from the y-coordinate of the point on the new graph. In general terms, we are investigating different values for k in the general equation $y = x^2 + k$. The value of k tells the amount of **vertical shift** that will translate the parent function $y = x^2$ into the new function $y = x^2 + k$. This is a **shape preserving** translation; the new graph has exactly the same shape as the parent function, but is in a different position in the coordinate plane. (Was the previous example of $y = ax^2$ for various values of a a shape preserving operation?)

Use the **DrawF** command from the DRAW menu to add many more graphs of the form $y = x^2 + k$ to the graphics screen by entering different values for k. How does the vertex of each parabola relate to the value of k? Explain.

What will happen if the coefficient, a, is negative? For example, draw the graph of $y = -x^2 + 1$. Before drawing the graph, predict the vertex, shape, and position of the graph. We know that if $k = 1$, the graph will be shifted upward 1 unit. Since $a = -1$, the graph should have the same shape as $y = x^2$, but open downward. If we start with the parent function $y = x^2$, reflect it across the x-axis, and then shift the graph upward 1 unit, the vertex of the new graph will be (0, 1) with the graph opening downward. Figure 2.54 shows that our analysis was correct.

What is the difference between the graph of $y = -x^2 + 1$ and the graph of $y = -(x^2 + 1)$? Can you predict the shape, position, and vertex of this graph? Explain how the parent graph of $y = x^2$ is transformed into the graph of $y = -(x^2 + 1)$. Figure 2.55 shows this graph. ◊

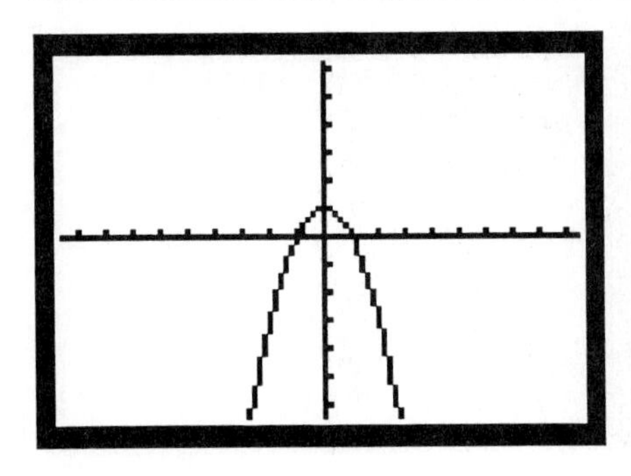

FIGURE 2.54
Graph of $y = -x^2 + 1$

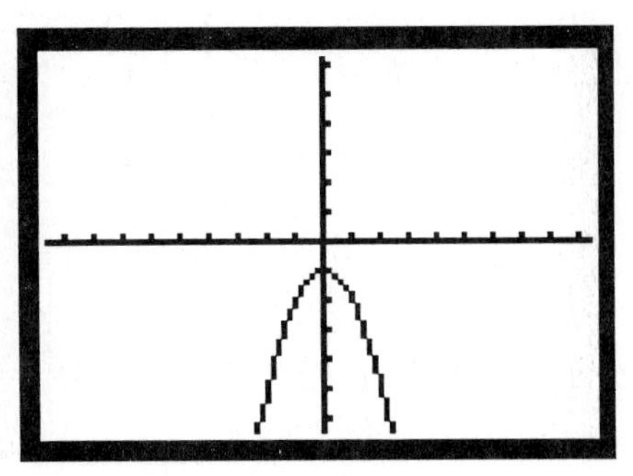

FIGURE 2.55
Graph of $y = -(x^2 + 1)$

EXPLORATION 5: Draw the graphs of $y = x^2$ and $y = (x - 2)^2$ in the same viewing rectangle. Investigate the effect of changing the value of h in the general equation $y = (x - h)^2$.

Figure 2.56 shows the graphics screen with the two graphs drawn. What is the vertex of the graph of $y = (x - 2)^2$? Add the graphs of $y = (x - 4)^2$ and $y = (x - 5)^2$ (see Fig. 2.57). What are the coordinates of the vertices of each of these graphs?

It appears that a pattern is forming. When $h = 2$, the graph of $y = (x - 2)^2$ is shifted 2 units to the right. The vertex of $y = (x - 2)^2$ is (2, 0). Likewise for the other graphs: The vertex of $y = (x - 4)^2$ is (4, 0), and the vertex of $y = (x - 5)^2$ is (5, 0).

Predict what will happen if $h = -5$ in the equation $y = (x - (-5))^2$ (see Fig. 2.58). You may enter this equation in the form $y = (x - (-5))^2$ or $y = (x + 5)^2$. Note the difference between the subtraction sign "–" and the opposite of a number sign, "-."

Use the **DrawF** command to add more members of the family of functions of

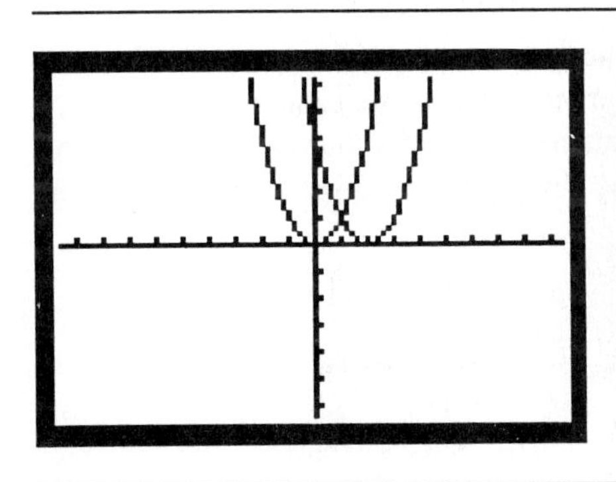

FIGURE 2.56

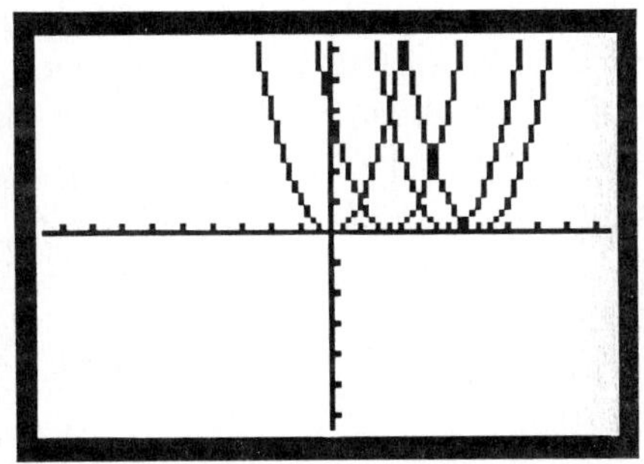

FIGURE 2.57

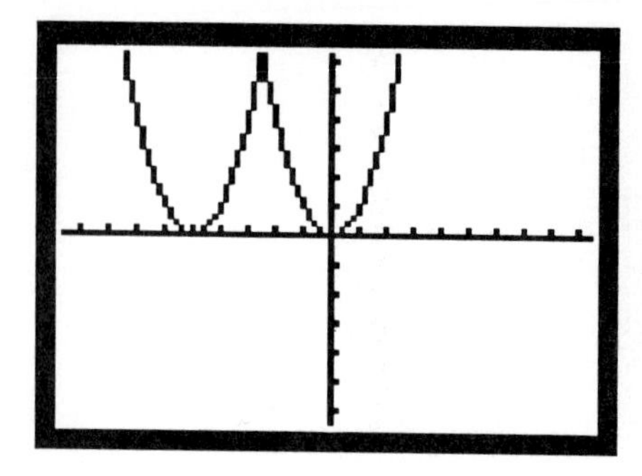

FIGURE 2.58

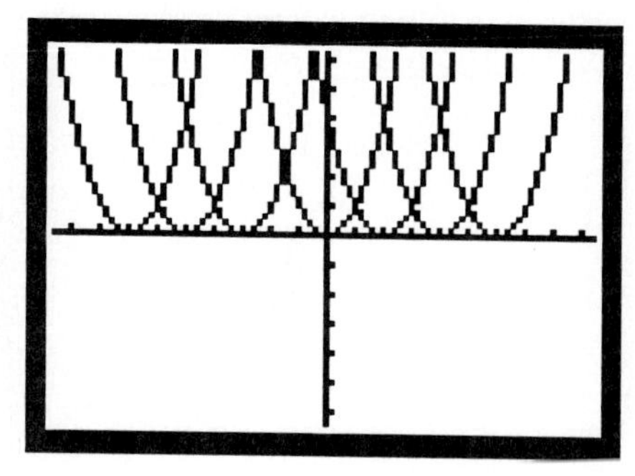

FIGURE 2.59

the form $y = (x - h)^2$ to the graphics screen. Figure 2.59 shows the screen with many graphs added. Can you make a general rule for the relationship between the equation $y = (x - h)^2$, the value of h, the coordinates of the vertex, the shape of the graph, and its position in the coordinate plane?

Now that we have explored the various types of stretches, shifts, and flips that can be done to the graph of a quadratic function, make a prediction about the shape, position, and vertex of the graph of $y = -3(x + 2)2 + 1$. Draw a graph of this quadratic function to confirm your prediction. Can you make a general rule to cover any general quadratic function in the form $y = a(x - h)^2 + k$? ◊

FOR ADVANCED ALGEBRA STUDENTS

EXPLORATION 6: Change the following quadratic function in general form to a quadratic function in the form $y = a\,(x - h)^2 + k$:

$$y = -3x^2 - 6x + 1.$$

The algebraic process used to change the general form of a quadratic into the $y = a(x - h)^2 + k$ form is called **completing the square**. If you examine this form of a quadratic equation, you will see that it contains the quantity "$(x - h)^2$," which is the square referred to in the name of the procedure. The following is one method for completing the square:

$y = -3x^2 - 6x + 1$	original general form
$\frac{y}{-3} = x^2 + 2x - \frac{1}{3}$	divide both sides by -3 to make the coefficient of $x^2 = 1$

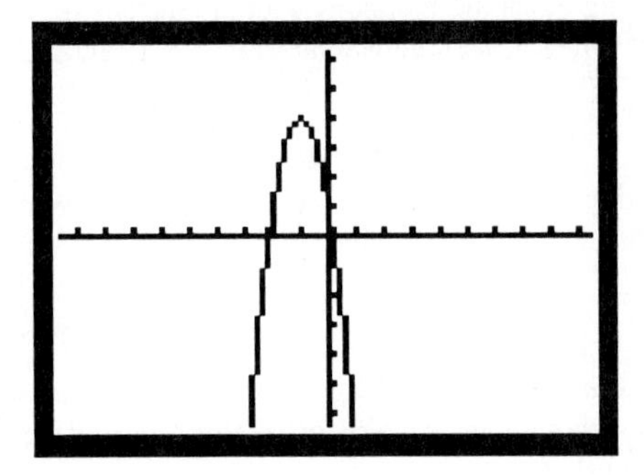

FIGURE 2.60

$\frac{y}{-3} = (x^2 + 2x + 1) - \frac{1}{3} - 1$	add and subtract the square of half the coefficient of the x term (b in the general term)
$\frac{y}{-3} = (x + 1)^2 - \frac{1}{3} - 1$	factor the perfect square term
$y = -3(x + 1)^2 + 1 + 3$	multiply through by -3
$y = -3(x + 1)^2 + 4$	collect like terms; finished form

In this form, $a = -3$, $h = -1$, and $k = 4$. The coordinates of the vertex are (-1, 4); the equation of the line of symmetry is $x = -1$; the parabola opens downward and is stretched by a factor of 3. Could you sketch this graph based on this information? Confirm your prediction on the TI-81 by graphing. Use the TRACE function to see numerical information on the screen. Figure 2.60 shows the graph of this parabola in the viewing rectangle [-9.6, 9.4] by [-6.4, 6.2].

If you were given a quadratic function in the form $y = a(x - h)^2 + k$, could you convert it to a quadratic in general form $y = ax^2 + bx + c$? What is the relationship between $a, b,$ and c and $a, h,$ and k? (*Hint*: Complete the square on the general quadratic $y = ax^2 + bx + c$.) Could you write a program for the TI-81 that would convert between the two forms of the equations?

Problems

1. Compare the following quadratic functions with the **parent function**, $y = x^2$. Predict the effect of changing a in the general equation $y = ax^2$. Draw the graphs to confirm your conjectures.

   a. $y = 3x^2$ **b.** $y = -5x^2$ **c.** $y = .5x^2$ **d.** $y = -.75x^2$

2. Compare the following quadratic functions with the **parent function,** $y = x^2$. Predict the effect of changing k in the general equation $y = x^2 + k$. Draw the graphs to confirm your conjectures.

 a. $y = x^2 + 1$ b. $y = x^2 - 1$ c. $y = x^2 + 2$ d. $y = x^2 - 3$

3. Compare the following quadratic functions with the **parent function,** $y = x^2$. Predict the effect of changing h in the general equation $y = (x - h)^2$. Draw the graphs to confirm your conjectures.

 a. $y = (x + 1)^2$ b. $y = (x - 1)^2$ c. $y = (x - 3)^2$ d. $y = (x + .5)^2$

4. Use the **parent function** $y = x^2$ as a starting reference and sketch the graph of the following quadratic functions without your graphing calculator. Confirm your sketches by graphing these curves on your calculator.

 a. $y = -3x^2$ b. $y = 3x^2 - 2$ c. $y = -3(x - 2)^2$
 d. $y = -(x + 1)^2 + 1$ e. $y = 3(x - 2)^2 + 1$ f. $y = -(x + 2)^2 - 3$

5. We have graphically investigated the transformation of the function $y = (x - h)^2$ compared with the function $y = x^2$.

 a. Make a conjecture about the graph of $y = (h - x)^2$ compared with the graph of $y = (x - h)^2$. Confirm your conjecture for several values of h, and explain what happens.

 b. Make a conjecture about the graph of $y = (h + x)^2$ compared with the graph of $y = (x + h)^2$. Confirm your conjecture for several values of h. Compare this to the results from part (a).

6. Graphically investigate the effect of the changing values for a, b, and c in the general quadratic function $y = ax^2 + bx + c$ by changing them as indicated in the following. Make a general rule for each of the three coefficients. Be sure to include negative and fractional values in your investigation.

 a. Let $a = 1$, $b = 1$, and change c.

 b. Let $b = 1$, $c = 1$, and change a.

 c. Let $a = 1$, $c = 1$, and change b.

7. For a general quadratic function $y = ax^2 + bx + c$, make and test conjectures about the effects of the following changes. To investigate these changes, enter a quadratic function of your choice in $\mathbf{Y_1}$, and the changed version of the same function in $\mathbf{Y_2}$; graph both functions in **Sequence** mode. Check your hypothesis with several different quadratics, and make a general rule.

 a. Change a to its reciprocal.

 b. Change a from positive to negative.

 c. Change b from positive to negative.

 d. Change c from positive to negative.

8. Experiment with different values of a, b, or c to solve each of the following. There may be more than one solution.

 a. Find b if $x^2 + bx - 2 = 0$, and 1 is a root.

 b. Find b to make the vertex of $y = x^2 + bx + 1$ closest to the origin.

 c. Find c if $y = 4x^2 + 4x + c$ touches the x-axis only once.

 d. Find a if $y = ax^2 + 8x + 2$ touches the x-axis only once.

e. Find a if $y = ax^2 + x + 1$ has two roots.

f. Find b if $y = x^2 + bx + 3$ has two roots.

9. Graph the following quadratic functions, and make a conjecture about the position of the graph when $c = \left(\frac{b}{2}\right)^2$ and $a = 1$. (*Hint*: Factor the quadratics; what form do they seem to take?) Confirm your conjecture on other quadratics. Can you explain what happens in terms of a transformation?

a. $y = x^2 + 2x + 1$ **b.** $y = x^2 + 4x + 4$ **c.** $y = x^2 + 6x + 9$

10. Graph $y = x$ in the window [0, 95] by [-3, 60]. In the interactive mode or from the **home screen,** draw a line from the point (5, 5) on the line $y = x$ to the point (60, 0) on the x-axis. Next, draw a line from the point (10, 10) on the line $y = x$ to the point (55, 0) on the x-axis. Continue this pattern every 5 units until you have connected the point (60, 60) on the line $y = x$ to the point (5, 0) on the x-axis. What shape is outlined by these lines? Explain.

11. Given that the roots are -3 and 5, write the quadratic function and check it by graphing. Is there only one solution? Suppose the additional condition is added that the y-intercept is 2? How many solutions are there under these conditions?

12. While a carpet was being installed in a square living room, a neighbor asked for an estimate for her living room. The installer said that since the neighbor's living room was twice the length and width, it would double the cost. The neighbor quickly asked to get this in writing. Why? Examine the graphs of $y = x^2$, $y = 2x^2$, and $y = (2x)^2$ using the TRACE cursor. Switch from one graph to the other to explain why the neighbor is getting such a good deal. How does this problem change if the side length of the living room triples, quadruples, etc.?

13. The area of the ring formed by two concentric circles where the larger circle has twice the radius of the smaller circle is triple the area of the smaller circle (see Fig. 2.61). Examine the area of this figure graphically for different radii; then prove this algebraically. (*Hint*: Let x stand for the radius, r.)

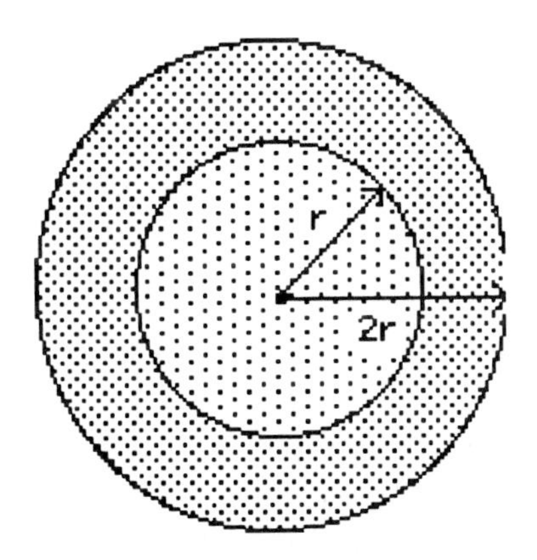

FIGURE 2.61

14. Find the legs and hypotenuse of a right triangle if their lengths are consecutive integers. Find these sides if their lengths each differ by 2. (*Hint*: Remember the Pythagorean Theorem, $a^2 + b^2 = c^2$.) What if the lengths of the sides differ by 3, 4, or 5? Is there a pattern?

15. Find two consecutive integers such that four times the first is equal to the square of the second. Write an equation to represent this problem situation, and sketch the graphs of both sides of the equation to estimate a solution before you use your TI-81. Solve the problem by graphing both sides of the equation. Check your solution both algebraically and numerically.

16. One very common mistake in algebra is to conclude that $(x + 3)^2 = x^2 + 9$. Show this is not true by graphing both sides of the equation separately. Add the graph of the correct expansion of $(x + 3)^2$ to confirm your findings.

17. Let x represent integer values. Then the expressions $2x$, $2x + 2$, $2x + 4$, . . . represent even integers and the expressions $2x + 1$, $2x + 3$, $2x + 5$, . . . represent odd integers. Graphically investigate the following products in the viewing window [-48, 47] by [-3200, 3100], and use the TRACE cursor to help make conclusions about the nature of these products.

a. The product of two even integers.

b. The product of two odd integers.

c. The product of an even integer and an odd integer.

18. The famous mathematician Karl Friedrich Gauss (1777–1855) discovered the formula for the sum of consecutive whole numbers as a result of a "busy work" assignment by his teacher when he was a young student. He was given the assignment to add all the whole numbers from 1 to 100. After a very short period of time he returned to the teacher's desk with the correct answer, 5050. He had seen the following pattern by writing the sum of the numbers forward and backward and adding each of the terms:

$$
\begin{array}{rcccccccccccccc}
S = & 1 & + & 2 & + & 3 & + & 4 & + & 5 & + & 6 & + & \cdots \\
S = & 100 & + & 99 & + & 98 & + & 97 & + & 96 & + & 95 & + & \cdots \\
\hline
2S = & 101 & + & 101 & + & 101 & + & 101 & + & 101 & + & 101 & + & \cdots
\end{array}
$$

Since there are 100 terms in this series, $2S = 100(101)$ and $S = \dfrac{100\,(101)}{2} = 5050$. If you let x represent the largest number in the series, then the sum of the first x whole numbers starting at 1 is given by the formula $S_x = \dfrac{x\,(x + 1)}{2}$. Graph Gauss's formula in a friendly integer window, and use the TRACE cursor to convince yourself that this works.

FOR ADVANCED ALGEBRA STUDENTS

19. Change the following to the general quadratic form, $y = ax^2 + bx + c$, and check your answers by graphing both forms.

a. $y = 3(x + 2)^2 - 4$ **b.** $y = -2(x - 5)^2 + 3$

20. Change the following from the general quadratic form, $y = ax^2 + bx + c$, to the completed square form, $y = a(x - h)^2 + k$, using the completing the square process. Check your answer by graphing both forms. Find the coordinates of the vertex of each parabola graphically and algebraically.

a. $y = x^2 + 6x - 5$ **b.** $y = 5x^2 - 10x + 1$

c. $y = -6x^2 + 18x - 2$ **d.** $y = -3x^2 - 7x - 5$

21. For the following quadratic functions, graph both $f(k + x)$ and $f(k - x)$ for $k = 1$. What happened to the graph? Try several other values of k. Try other quadratic functions. Express what happened using the idea of symmetry, and explain it algebraically.

a. $f(x) = x^2 + 5x + 6$ **b.** $f(x) = 1 - 3x - 2x^2$

22. Graph the quadratic function $y = x^2$ and the line $y = 2x - 1$. What do you notice? Try the quadratic function $y = 4x^2$ and the line $y = x - 1$. For any quadratic function $y = \frac{x^2}{4p}$, where p is the distance from the vertex to the focus, the tangent line at the point (a, b) on the curve has $-b$ as its y-intercept, and its equation is $y = \frac{a}{2p}x - b$. Check this for the following quadratic functions.

a. $y = \frac{x^2}{8}$ at $(4, 2)$ **b.** $y = \frac{1}{2}x^2$ at $(4, 8)$ **c.** $x^2 = 4y$ at $(4, 4)$

23. Using the equation given in problem 22, graph the two tangent lines for the following quadratics through the given symmetrical points on the curves. If you are in a friendly window, you should be able to make a conjecture about the point of intersection of the two tangent lines and be able to solve for the point algebraically. (*Hint*: Check the slopes.) Can you make a general rule about the point of intersection?

a. $y = 4x^2$ and points $(1, 4)$ and $(-1, 4)$

b. $y = -x^2$ and points $(2, 4)$ and $(-2, 4)$

24. Graph the following and explain the results. Write and check several of your own functions with similar behavior.

a. $y = \sqrt{x^2 + 2x + 1}$ **b.** $y = \sqrt{x^2 + 4x + 4}$ **c.** $y = \sqrt{x^2 - 6 + 9}$

25. Use the reciprocal key ($\boxed{X^{-1}}$) to graph and investigate the following functions. Make and test a conjecture about the behavior of the graphs as a result of the number of roots in the denominators of these rational functions.

a. $y = \frac{1}{x^2 + x + 1}$ **b.** $y = \frac{1}{1 + x^2}$ **c.** $y = \frac{1}{-2x^2 + 2x - 1}$

26. Investigate when $(x^2 + x + 1)^{(x^2 + 3x + 2)} = 1$, and explain your answers algebraically. Enter the base and exponent as separate functions, and then combine the two parts in $\mathbf{Y_3}$ (see Fig. 2.62). Activate $\mathbf{Y_1}$, $\mathbf{Y_2}$, and $\mathbf{Y_3}$ separately to investigate the behavior of the base, the exponent, and the function. (*Hint:* Remember your basic exponent rules.)

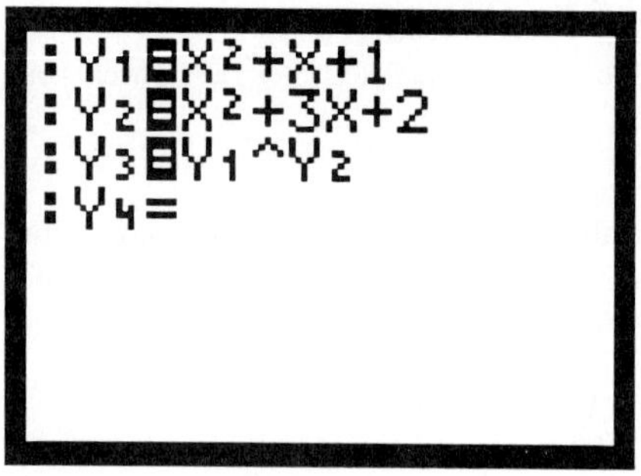

```
:Y1=X²+X+1
:Y2=X²+3X+2
:Y3=Y1^Y2
:Y4=
```

FIGURE 2.62

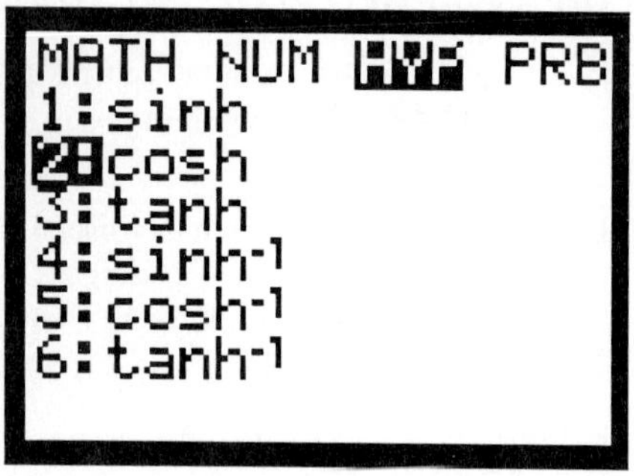

```
MATH NUM HYP PRB
1:sinh
2:cosh
3:tanh
4:sinh-1
5:cosh-1
6:tanh-1
```

FIGURE 2.63

27. The equation $y = 3 \cosh \frac{x}{3}$ graphs a **catenary curve.** This is the curve formed by a free hanging cord or cable of uniform size attached at both ends, like a telephone cable. At first glance this shape looks like a parabola, and Galileo believed it was this shape. Convince yourself that a catenary curve is not a parabola by graphing it along with a parabola that closely resembles its shape. Use what you have learned in this section to find such a parabola.

The function "cosh" stands for the hyperbolic cosine function. This function can be found on the **HYP** submenu of the MATH menu as command **2: cosh** (see Fig. 2.63). Use the friendly viewing window [-9.6, 9.4] by [0, 12.3] to begin your experiment. When weight is applied to the catenary curve at equal intervals, it does become a parabola. This can be seen in cable suspension bridges like the Golden Gate Bridge. This will become evident as you experiment with different parabolas in an attempt to get the catenary curve.

28. Solve the general quadratic form $y = ax^2 + bx + c$ by completing the square to get an equation in the form $y = a(x - h)^2 + k$, show that $h = \frac{-b}{2a}$ and $k = c - \frac{b^2}{4a}$, and explore this relationship graphically for different values of a, b, and c.

29. The quadratic formula, $x = \frac{-b \pm \sqrt{b^2 - 4ac}}{2a}$, is an efficient way to find the roots of any quadratic function. If you know the values for the coefficients a, b, and c, then you can calculate the roots. Write a program for the TI-81 that will calculate the roots of a quadratic function based on the input of values for a, b, and c.

Extra Challenge: Modify your program so that it gives the real and imaginary parts for complex roots.

30. The quadratic form $y = a(x - h)^2 + k$ gives you information about the vertical stretch and horizontal and vertical shift needed to transform the parent function $y = x^2$ into any given quadratic function. The values of h and k are also the coordinates of the vertex of the transformed parabola. Use your results from problem 28 to expand the program you wrote for problem 29 so it will give information about the coordinates of the vertex of the transformed parabola and values for the completed square form of the quadratic, and will draw the graph of the function.

Extra Challenge: Modify your program so that an appropriate viewing window showing a complete graph of the quadratic function is automatically set from within the program.

2.3 Exploring Problems Represented by Quadratic Functions

Problem situations represented by quadratic functions provide the setting for a wide variety of questions and explorations. In this section we will explore several problems modeled by quadratic functions to investigate equations, inequalities, and extreme value problems.

EXPLORATION 1: Many rectangles can be constructed or drawn such that the length is equal to twice the width plus three. Investigate how the area of these rectangles varies as the width varies between zero and 16.

The problem asks us to investigate the area of the rectangles in terms of the width. For a rectangle, *area* = *width* × *length.* The following table is a numerical representation of the problem.

Notice that once we choose the width, we can calculate the length and the area of a given rectangle. If we pick a different width, we get a different length and area. We say that **area is a function of the width**, since the area of each rectangle depends on what we pick for the width. Length is also a function of the width for the same reason. But since we are investigating area as it relates to width, we will represent length in terms of width in the area formula: $l = 2w + 3$, and then $A = w(2w + 3)$. The last line in the table is a general statement that represents all other possible lines in the table. We call this the **formula**, or equation, of the problem situation.

To graph this equation on the TI-81, we will let **X** represent the independent variable in the equation, the *width* of the rectangles. The dependent variable, $\mathbf{Y_1}$, will

TABLE 2.3

Width	*Length*	*Area* = $w \times l$
0	$2(0) + 3 = 3$	$0 \times 3 = 0$
1	$2(1) + 3 = 5$	$1 \times 5 = 5$
5	$2(5) + 3 = 13$	$5 \times 13 = 65$
8	$2(8) + 3 = 19$	$8 \times 19 = 152$
13	$2(13) + 3 = 29$	$13 \times 29 = 377$
16	$2(16) + 3 = 35$	$16 \times 35 = 560$
x	$2x + 3$	$A = x(2x + 3)$

represent the *area* of the rectangles. Sometimes $\mathbf{Y_1}$ is also referred to as the function value. Enter the equation on the Y= menu as $\mathbf{Y_1 = X(2X + 3)}$ or $\mathbf{Y_1 = 2X^2 + 3X}$.

Draw the graph in an appropriate viewing rectangle. Since *width* is the independent variable and is in the range $0 \le w \le 16$, set **Xmin = 0** and **Xmax = 19** to get a friendly step of .2 in the horizontal direction (why?). If you look at the table, the range of values for *area* is $0 \le area \le 560$. Set **Ymin = 0** and **Ymax = 600**. We do not need a friendly step in the vertical direction since we will use the TRACE cursor, which returns a calculated *y* value and not a screen coordinate. Figure 2.64 shows the resulting graph.

Activate the TRACE cursor and check the values in the table. Each point that the TRACE cursor evaluates on the screen represents a line in a table. We did not include most of these lines in Table 2.3 because of space. But, you can think of the TRACE cursor as moving through a very large table of values as it slides along the function. Because the TRACE cursor will follow the function even off the screen, the number of lines represented in your table is **very large**. Remember, even if the TRACE cursor is not visible, the values at the bottom of the screen are accurate representations of *x*, the independent variable, and *y*, the calculated value of the function ($f(x)$) at the given value for *x*.

What happens to the area as the width of the rectangles increases beyond 16? (The values increase.) Why? Does this make sense in terms of the problem? As the width gets larger and larger, the length gets larger and the product of width and length also increases. What is the area of this type of rectangle whose width is 30 units? What is its length? Figure 2.65 shows the TRACE cursor returning a value for area based on a width of 30. The cursor is off the screen and the graph has scrolled to the right.

What happens if the width is allowed to get smaller than zero? This makes no sense in the **problem situation,** since the width must be a positive number. However, in the **algebraic representation,** the value of the independent variable could be negative. If you TRACE to the left so that the values of *x* become negative, what

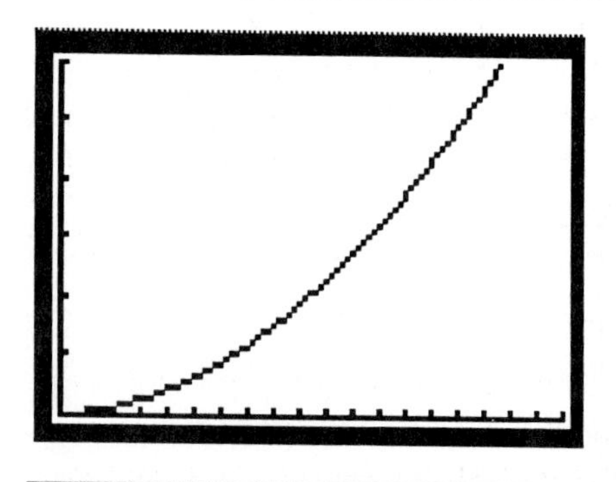

FIGURE 2.64

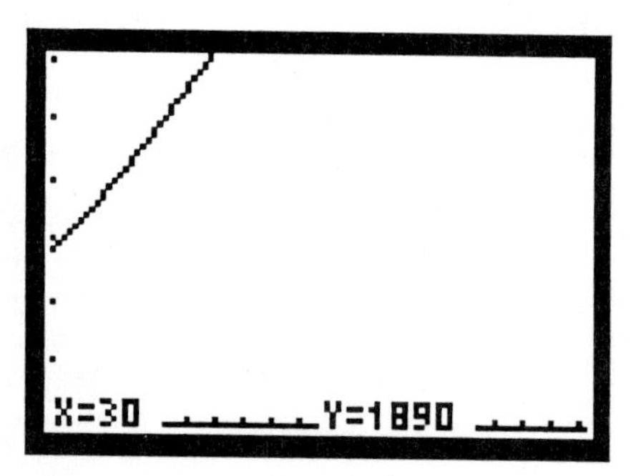

FIGURE 2.65

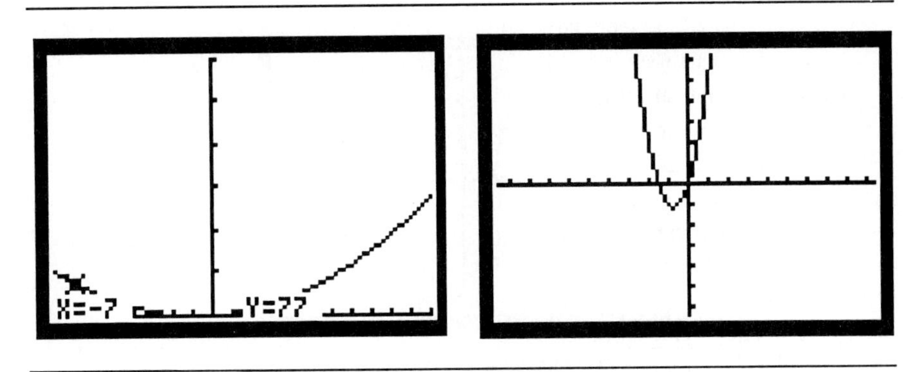

FIGURE 2.66

FIGURE 2.67

happens to the function values? Figure 2.66 shows the screen with the graph scrolled to the left. The function values are increasing. What should the shape of this graph be? Should the function values increase as x gets smaller (i.e. more negative)?

Figure 2.67 shows the same graph drawn in the viewing rectangle [-9.6, 9.4] by [6.4, 6.2]. The portion of the graph seen in Fig. 2.64 representing the problem situation is the branch appearing in the **first quadrant** of this graph. ◊

MAXIMUM AND MINIMUM VALUE PROBLEMS

In previous sections we graphed many parabolas representing quadratic functions. Each of these parabolas had a vertex, which represented the lowest or highest vertical point on the parabola. When the parabola opened upward, the value of a was positive and the graph had a lowest point. Sometimes we refer to this type of parabola as a **cup** since it resembles a cup for holding liquids. When the parabola opens upward, it has an **absolute minimum** value for the function in the vertical direction. When the value of a is negative, the parabola opens downward. This is called a **cap,** and the parabola has an **absolute maximum** value for the function in the vertical direction. Figure 2.68 shows the two parabolas $y = -2\ (x + 3)^2 + 4$ and $y = (x - 4)^2 - 5$ graphed in the viewing rectangle [-9.6, 9.4] by [-6.4, 9.2]. Draw these graphs on your TI-81, and use the TRACE cursor to check the maximum or minimum value for each parabola. Do these points correspond to the vertices of the two parabolas? The vertex is also known as the **turning point** of the parabola, because the vertical direction of the graph reverses at this point.

EXPLORATION 2: Tim McGrath has a very productive garden, but he also has many very hungry rabbits for neighbors. He purchased a 320-foot roll of rabbit-proof fence to put around the garden. His garden is situated next to a stone wall, so he only needs to put the rabbit-proof fence on three sides of his rectangular garden. Tim would like

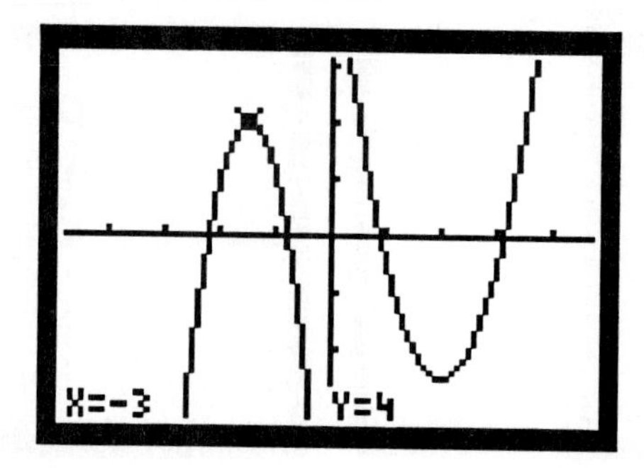

FIGURE 2.68

to enclose the largest rectangular plot of land possible within the fence. What are the dimensions and the area of the largest garden Tim can protect from the rabbits?

Figure 2.69 shows a drawing of the problem situation. Let x represent the width of the garden. Since the total length of the fence is 320 feet, the length of the garden is given by $l = 320 - 2x$. The area, A, of the garden is given by $A = lw$, or $A = x(320 - 2x)$, or $A = 320x - 2x^2$. Area is a function of the width. Enter either equation on the [Y=] menu: $\mathbf{Y_1 = X(320 - 2X)}$ or $\mathbf{Y_1 = 320X - 2X^2}$.

The "guess and check" method of finding an appropriate range for this graph is not the best or most efficient way to set the viewing window. Instead, let's look at the problem itself. The variable x represents the width of the garden. The smallest x can be is 0 and the largest is 160. (Why?) The function values (the y values) represent the area of the rectangular garden. The smallest the area of the garden can be is zero (not a very productive garden!). We are trying to find the largest value for the area of the

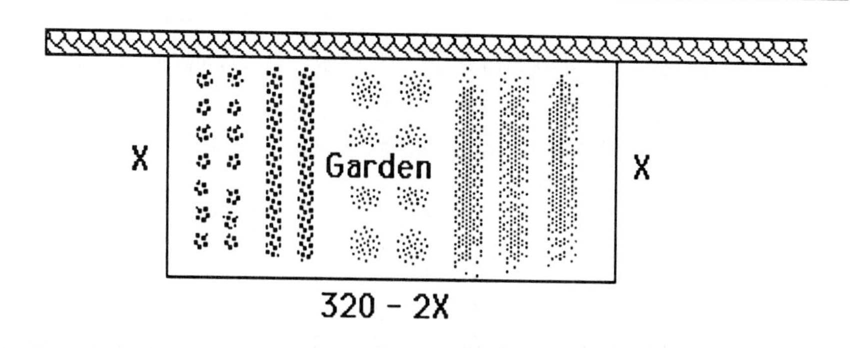

FIGURE 2.69
Garden adjacent to the stone wall

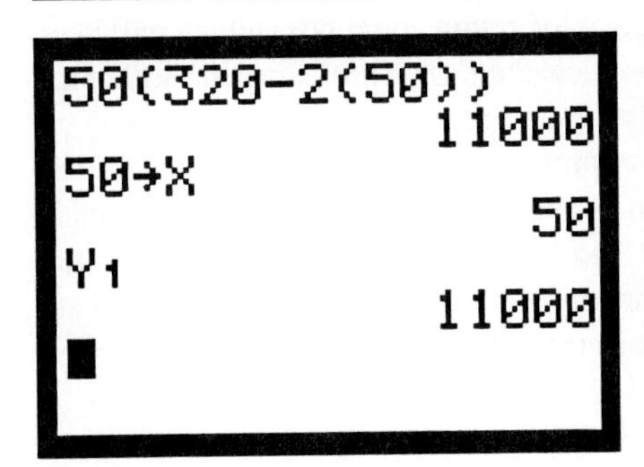

FIGURE 2.70

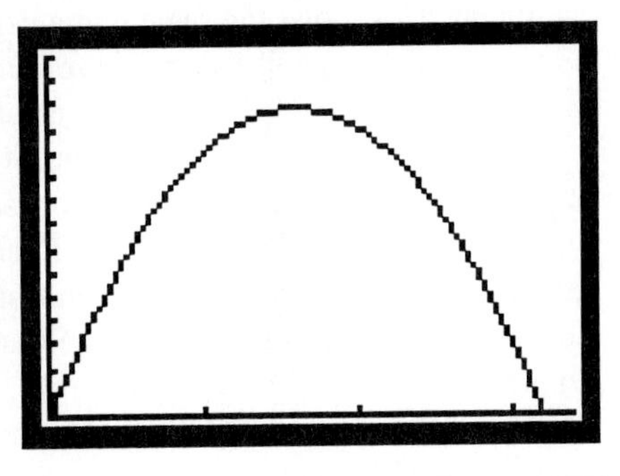

FIGURE 2.71

garden. Pick a value for the width, and find the area of the garden. Figure 2.70 shows the value 50 used for the width in the formula. The resulting area is 11,000 square feet. An alternate way to check function values is to store the value for the width in the variable **X** and evaluate $\mathbf{Y_1}$ as shown in Fig. 2.70.

By checking several other values for the width (for example, 60, 70, or 75), it is clear that the maximum value for the area must be above 12,000 square feet. Set the viewing window of the TI-81 at [0, 170] by [0, 15,000]. Figure 2.71 shows the graph in this window. Use the TRACE cursor to investigate how the area changes as the value of x, representing the width, changes from 0 to 170 feet. The vertex of the graph, a maximum value in this problem, seems to be somewhere near the point (80.5, 12,799) (see Fig. 2.71).

To get a better estimate of the maximum value, use the **Box** command from the ZOOM menu to zoom in on the graph. Figure 2.72 shows the box in position for the first zoom. If you repeat this type of square zoom box several times, the graph will look like a straight line, as shown in Fig. 2.73.

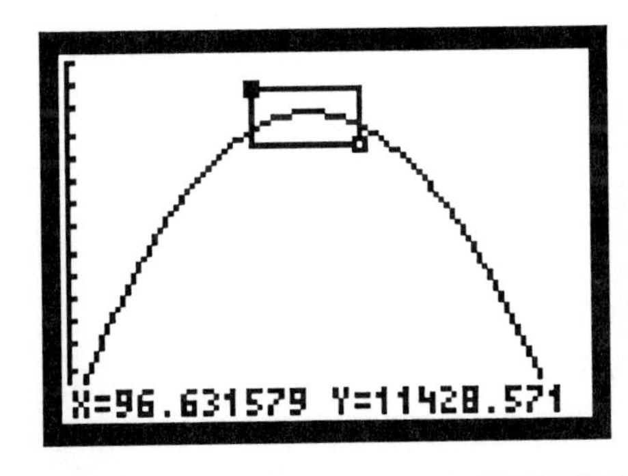

FIGURE 2.72

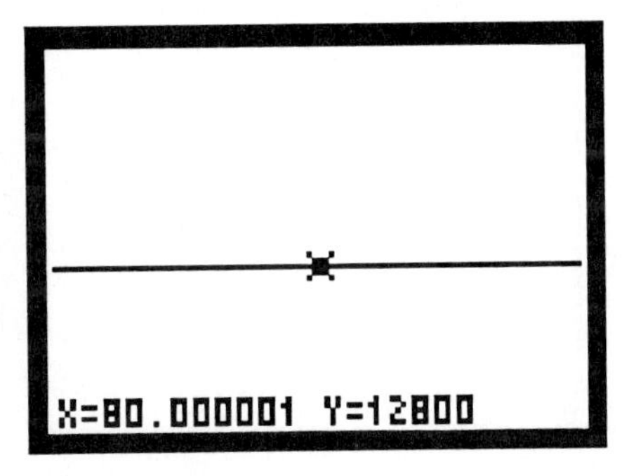

FIGURE 2.73

Figure 2.74 shows the same graph using a long, thin zoom box, which will keep the curvature in the graph by stretching the vertical dimension more than the horizontal dimension. Figure 2.75 shows the graph after zooming.

This technique is called **differential zooming,** because it zooms more in one direction than the other. The same effect can be obtained using the **Zoom In** command by setting different factors for the **XFact** and **YFact**. For example, select the **Set Factors** option from the ZOOM menu, and let **XFact = 5** and **YFact = 10**. Then use the **Zoom In** command to zoom to a point selected on the screen by the moving **screen cursor**.

Note: When using a differential zoom procedure, it usually takes more steps to reach a desired level of accuracy in both the horizontal and vertical direction because one dimension is zooming in faster than the other.

After several zoom operations by any method you choose, use the TRACE cursor to estimate the value of the maximum point of the graph. Since we are investigating a quadratic function, this point will be the vertex of the parabola. Figure 2.76 shows an estimate of this point after several zooms.

With error at most .01 unit for both the width and area, the maximum area of the garden is 12,800 square feet, when the width of the garden is 80 feet. By calculation, the length of the garden will be 320 – 2(80) = 160 feet. Figure 2.77 shows two methods by which the width of 80 can be used to evaluate the function. An area of 12,800 square feet is returned as the function value in both cases. Tim can now put up his fence and keep the rabbits out of his garden. ◊

EXPLORATION 3: The Quandary Game Company, Inc. has created a new hand held game system for kids of all ages. The research and development (R & D) costs for the new system followed a quadratic model with the equation $cost = -3x^2 + 550x + 50{,}000$, where x represents the number of units sold (in thousands).

(a) Graph the equation in an appropriate viewing window to show a complete graph.

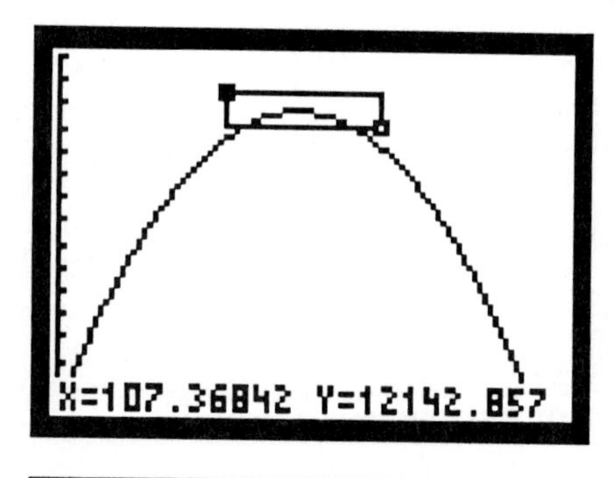

FIGURE 2.74

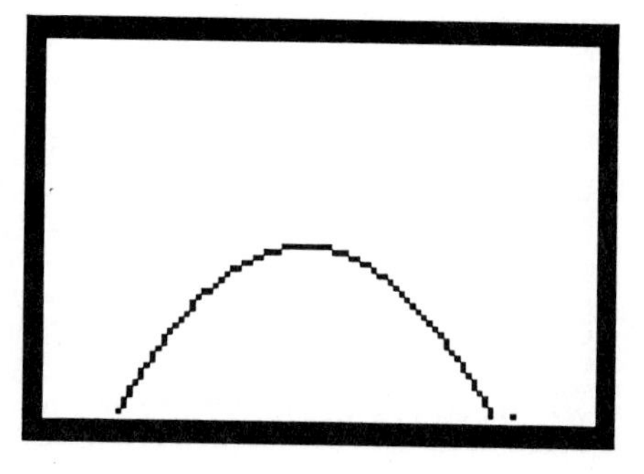

FIGURE 2.75

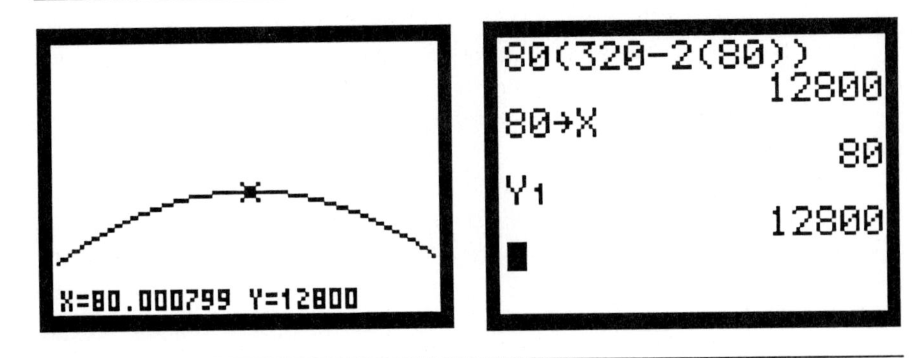

FIGURE 2.76

FIGURE 2.77

Enter the equation $\mathbf{Y_1 = -3X^2 + 550X + 50000}$ on the Y= menu. Store different values in the variable **X**, and evaluate $\mathbf{Y_1}$ to get an idea about the behavior of the graph. The value for the R & D costs (the function values) should be positive. Figures 2.78 and 2.79 show examples of this. Based on these calculations, it appears that the values for x, the number of units sold (in thousands), should be represented on the axis between -100 and 300. (Do negative values of x have any meaning in this problem setting?) The function's values seem to be less than 100,000. (Why do we have to go up to at least 50,000 to see anything important?) Set the viewing window [-100, 300] by [-10,000, 100,000]. Figure 2.80 shows the resulting graph.

Note: Often your first viewing window will not be exactly what you want. The window just specified was the result of minor modifications after the first try. Analyzing the function before attempting to graph the function gets you close to the appropriate viewing window. Modify your RANGE values and regraph as many times as needed to produce a complete, useful graph.

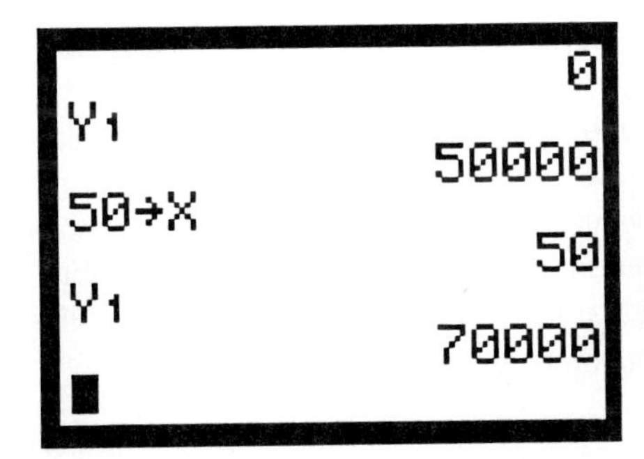

FIGURE 2.78

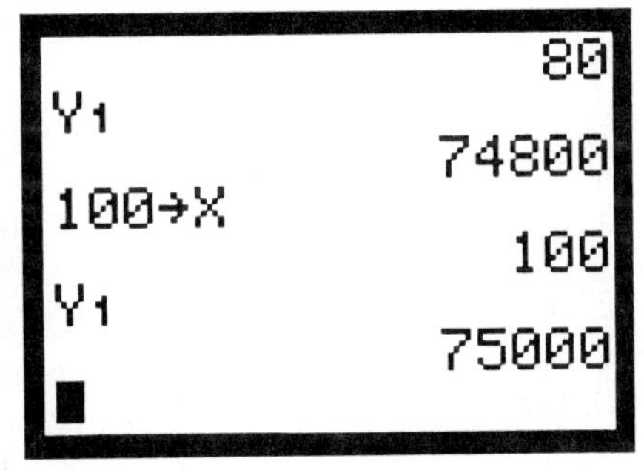

FIGURE 2.79

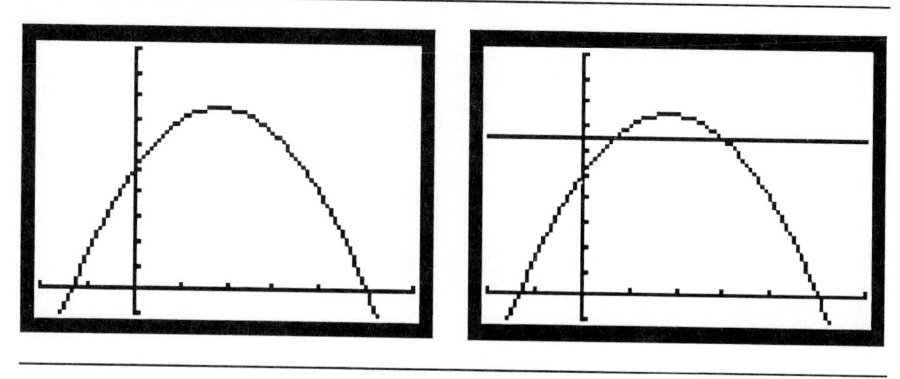

FIGURE 2.80

FIGURE 2.81

(b) Use the graph to explain what happened to the R & D costs in the time period shortly after the first units were offered for sale. Continue your analysis of the graph to explain what will eventually happen to the R & D costs.

The point where $x = 0$ (the y-axis) represents the point when the new system first went on sale. At that time, when 0 units had been sold, the R & D costs were \$50,000. During the first sales phase, the R & D costs rose to a maximum of approximately \$75,200 at $x = 90$, or when 90,000 units had been sold. After that maximum point, R & D costs dropped off steadily. They reached the same level as when the product was first introduced (\$50,000) after approximately 182,000 units had been sold ($x = 182$). When approximately 250,000 units are sold, R & D costs will be zero.

(c) Determine when, in terms of units sold, the R & D costs are greater than \$65,000.

The R & D costs will be greater than \$65,000 when $-3x^2 + 500x + 50{,}000 > 65{,}000$. To represent this inequality, graph both sides as functions:

$$\mathbf{Y_1 = -3X^2 + 550X + 50{,}000} \quad \text{and} \quad \mathbf{Y_2 = 65{,}000.}$$

Figure 2.81 shows the resulting graph in the window [-100, 300] by [-10,000, 100,000].

The solution to the inequality is represented by the interval on the x-axis where the graph of the function $\mathbf{Y_1 = -3X^2 + 550X + 50{,}000}$ is above the graph of the line $\mathbf{Y_2 = 65{,}000}$. Zoom in on the intersections of the line and the parabola to find the ends of the interval representing the solution of the inequality. With error at most .01, the interval $\mathbf{33.33 < x < 150}$ represents the solution set of the inequality. Figure 2.82 shows these endpoints used to evaluate the function.

(d) Determine when, in terms of units sold, R & D costs drop below the level they were when the first units were sold to the public.

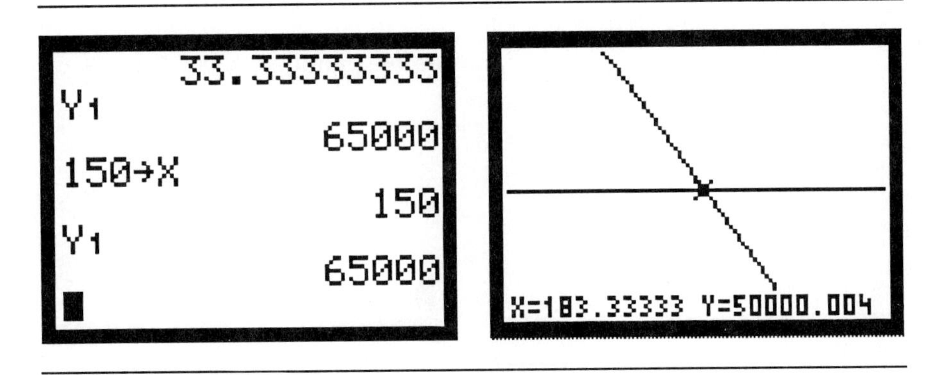

FIGURE 2.82

FIGURE 2.83

Solve the inequality $-3x^2 + 500x + 50{,}000 \leq 50{,}000$ by graphing both sides and zooming. Figure 2.83 shows the point on the graph with error of at most .01. After 183,333 units are sold, the R & D costs will be less than or equal to \$50,000.

(e) What are the maximum R & D costs and when do these occur, in terms of units sold?

The maximum R & D costs are represented by the vertex of the parabola. Zoom in to find the coordinates of this point. Figure 2.84 shows this point. The maximum R & D costs are approximately \$75,208.33, after 91,667 units have been sold.

(f) When, in terms of units sold, will the R & D costs be completely covered by profit from the sale of the new game systems? Or, when are R & D costs zero?

All R & D costs are covered by profit from sales when the graph of R & D costs crosses the x-axis. This point corresponds to the solution to the quadratic equation

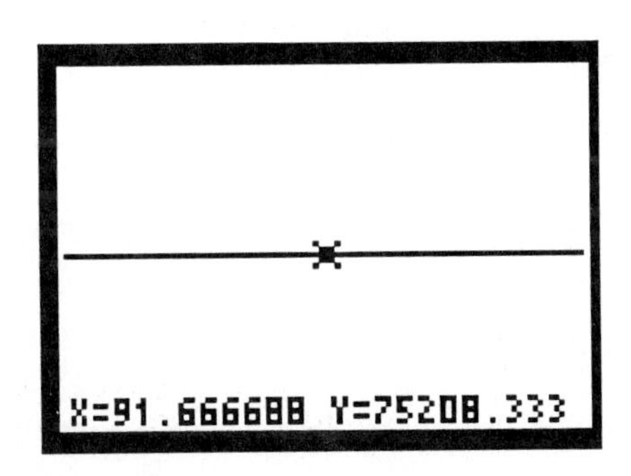

FIGURE 2.84

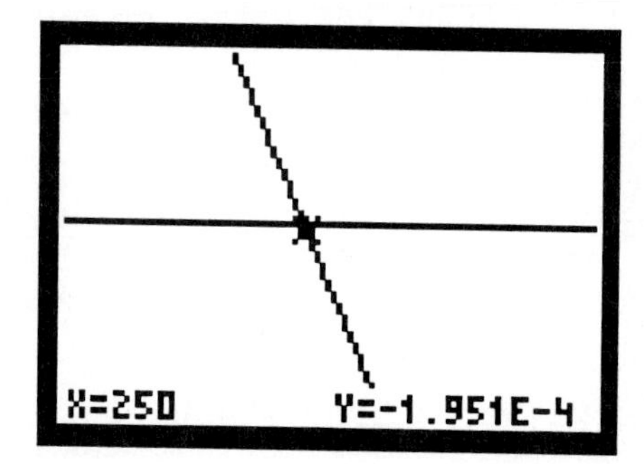

FIGURE 2.85

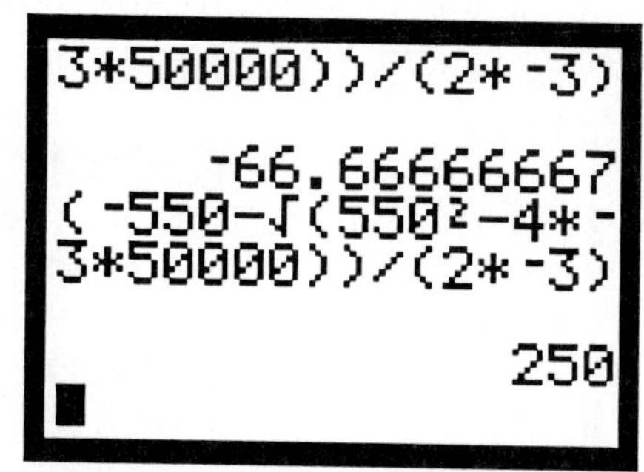

FIGURE 2.86

$-3x^2 + 500x + 50{,}000 = 0$. Zoom in on this point (see Fig. 2.85). R & D costs are zero when 250,000 units have been sold.

An alternate solution to this part of the question can be done by the quadratic formula. Figure 2.86 shows the calculation of the two roots of the equation based on the input $x = \dfrac{-550 \pm \sqrt{550^2 - 4(-3)(50{,}000)}}{2(-3)}$. The solutions are $x = -66.66666667$ and $x = 250$. The negative solution is not part of the problem situation; only the positive solution has meaning. This value represents 250,000 units sold and corresponds to the solution found graphically. ◊

Problems

1. Find the vertices (maximums or minimums) of the following quadratic functions with error at most .01 using the **Box** command from the ZOOM menu.

a. $x^2 - 4x + 1$ **b.** $6x^2 + 7x - 5$ **c.** $7 - 3x - 2x^2$

2. The x-coordinate of the vertex of a parabola is given by the formula $x = \dfrac{-b}{2a}$, where a and b are the coefficients from the general quadratic form $y = ax^2 + bx + c$. Graph the following quadratic functions, and use the TRACE cursor to confirm that the x-coordinate of the vertex is given by this formula. (*Note*: You may need to zoom in to check the value more closely.)

a. $y = x^2 + 4x + 2$ **b.** $y = 3x^2 - 4x + 5$ **c.** $y = 7 - x - x^2$

3. The y-coordinate of the vertex of a parabola can be calculated by finding the function value for the x-coordinate of the vertex. For example, if $x = \dfrac{-b}{2a}$ gives the x-coordinate of the vertex, then $f\left(\dfrac{-b}{2a}\right)$ gives the y-coordinate. Calculate the coordinates of the vertices of the following parabolas. Confirm by graphing that the vertex of each parabola is the point $\left(\frac{-b}{2a}, f\left(\frac{-b}{2a}\right)\right)$.

a. $y = x^2 + 3x + 2$ b. $y = 2x^2 - 4x - 7$ c. $y = 2(x + 4)^2 - 3$

4. According to one student, the x-coordinate of the vertex is halfway between the roots. Explain why you agree or disagree, and use the following to check your answer.

   a. $y = x^2 + 2x - 3$ b. $y = 5x^2 - x - 8$ c. $y = -x^2 + 5x - 5$

5. Consider all the different rectangles that have a perimeter of 80 cm.
   a. Draw a diagram representing the problem situation. Let x represent the width of the rectangles.
   b. Write an algebraic expression for the length in terms of the width.
   c. Write an equation for the area of the rectangles as a function of the width, x.
   d. Graph this function and find the maximum possible area for a rectangle with perimeter 80 cm.
   e. Confirm algebraically that the point you found in part (d) is the vertex.

6. Suppose you have two numbers whose sum is 100.
   a. Let x represent one number. Write an equation representing the second number in terms of x.
   b. Write an equation for the product of the two numbers as a function of x.
   c. Graph the function, and find the maximum product that could be formed by two numbers whose sum is 100.
   d. Repeat the same procedure for three other sums of your choosing, such as 150, 180, or 200. In each case find the combination of two numbers that produces the largest product.
   e. Make a generalization about the values that produce the maximum product and the size of the maximum product.
   f. Write a paragraph about the relationship between this problem situation and the rectangle problem situation appearing in problem 5.

7. When an object is thrown straight up into the air, the object's height is given by the formula $H(t) = -4.9t^2 + v_0t + s_0$, where t is the time, v_0 is the initial velocity, and s_0 is the initial height of the object. The height of the object is a function of the time the object has been traveling. If a ball was thrown vertically from ground level at a velocity of 85 meters per second, find the following:
   a. Write an algebraic expression representing the height as a function of time.
   b. Graph the function. Remember, let x represent the time, t, and y represent the height, $H(t)$.
   c. What values for x, the time of flight, make sense in the problem situation?
   d. When will the ball hit the ground?
   e. What is the maximum height, and when does it occur?
   f. When is the ball at least 300 meters above the ground?
   g. What if the ball was thrown from a building 30 meters high? Repeat parts (a)–(f) using this new situation. Repeat the problem again using a different starting height of your

choice. Graph up to four different situations at the same time. Compare your answers visually and numerically using the TRACE cursor for the different starting heights. What is the same for each different starting height? What is different? Make a conjecture about the effect of changing the starting height on the outcomes of the problem.

8. Joe Quigley has 200 feet of fence and wishes to enclose a rectangular plot of land adjacent to a large barn for a livestock paddock. He will use the side of the barn as one side of the enclosure. Find the dimensions and area of the rectangular paddock of maximum size.

9. As an alternative to one large paddock, Mr. Quigley is considering the option of using the 200 feet of fence to make two adjacent paddocks against the barn with a common side between them. Each of the two paddocks would be of equal size.

a. What are the dimensions and the total area of the two paddocks of maximum size?

b. Graph the problem situations in problems 8 and 9 on the same screen to compare the total area enclosed in two paddocks with the area enclosed in one large paddock.

c. What if the 200 feet of fence were used to make three equal paddocks, or four equal paddocks? Compare the graphs of all four problem situations at the same time.

d. Make a generalization about the dimensions of each individual paddock and the total amount of enclosed area for any number of paddocks.

10. A new zoning law in Isabella County specifies that residential building lots must have half the total area reserved as an undeveloped "green belt" of uniform width on all sides of the property. The house and other building cannot cover more than half of the total area of the lot. The Lee family purchased a rectangular lot 175 feet wide by 200 feet deep to build a house. Answer the following questions to help plan the new residence:

a. Draw a diagram of the problem situation.

b. Write an equation in terms of x, the width of the "green belt," that relates the building area to the actual area of half of the lot.

c. Represent the problem situation visually by graphing both sides of the equation from part (b).

d. Find the width of the green belt and the dimensions of the building area of the lot by zooming in on the graph. How many solutions are there? Explain.

e. Solve the problem algebraically using the quadratic formula to confirm your answer from part (d).

f. What if the law specified one-quarter or one-third of total area of the lot was to be a uniform green belt? Repeat the problem using these other possible sizes. Compare the graphs of all three situations on the same screen.

11. A rectangle is located in the coordinate plane such that it has one corner at the origin, and the opposite diagonal corner is located on the line $y = 3 - x$. Find the coordinates of the corner on the line such that the area of the rectangle is maximized.

12. Using TRACE and a friendly integer window, find two consecutive integers whose product is a six-digit number and where the:

a. second three digits of the product are equal to the first three digits of the product.

b. first three digits of the product are double the second three digits of the product.

c. second three digits of the product are double the first three digits of the product.

13. Graph $y = x^2$ and plot the point (2, 5). Which is smaller: (i) the distance from the point to the vertex of the parabola, or (ii) the distance from the point to the x-axis? Explain. Estimate the point on the graph of the parabola that is closest to the point (2, 5). Zooming with a friendly window will help get a good estimate. (Later, we will solve this problem more exactly.) What happened to the curve as you zoomed in? What kind of line measures the shortest distance between a point and a line?

FOR ADVANCED ALGEBRA STUDENTS

14. Graph $y = x^2 + bx + 1$ for different values of b. Try to find a value for b that makes the vertex closest to the origin. Use the fact that, in general, the vertex of a parabola is the point $\left(\frac{-b}{2a}, c - \frac{b^2}{4a}\right)$. Try graphing four different equations at once and zoom in.

15. A parabola is defined to be the set of points equidistant from a single point (called the focus) and a line (called the directrix). Use the **Pt-On** and **Line** commands to draw the focus and directrix of each of the following parabolas. For each parabola, use this definition to find three points that are on the parabola, and plot them. Write the equation of each parabola, and graph it using the **DrawF** command to see if each graph goes through all of its three points.

a. focus (1, 1) and directrix $y = -1$

b. focus (-1, -2) and directrix $y = 1$

c. focus (4, 6) and directrix $y = 8$

16. The following quadratics have no real roots. Find the complex roots $a + bi$ and $a - bi$ using the quadratic formula. Use the TRACE cursor to find the vertices of the quadratic functions. Do you notice anything about the coordinates of the vertices and the a and b values in the complex number $a + bi$?

a. $y = x^2 + 1$ **b.** $y = 2x^2 + 2x + 5$

c. $y = -x^2 + x - 1$ **d.** $y = 3 + 2x + .5x^2$

17. For the function $y = (x-a)^2 + (x-b)^2 + (x-c)^2$, the x-coordinate of the vertex is the average of the three values a, b, and c. Graph several equations in this form to convince yourself that this is true. Show your results numerically and prove this relationship algebraically.

18. For $y = c(x - a)^2 + d(x - b)^2$ the x-coordinate of the vertex is $x = \frac{ca + db}{c + d}$. Graph several equations in this form to convince yourself that this is true.

19. Find the minimum total area of an equilateral triangle and a square if their total perimeter is 10 and they have no sides in common.

20. Find the minimum total area of a circle and a square if their total perimeter is 16.

21. Graph $y = x^2$, and draw a triangle from the vertex and any two symmetrical points (x, y) and $(-x, y)$ on the quadratic.

a. Is the triangle isosceles? Prove it.

b. For what value of x would the symmetrical points and the vertex make an equilateral triangle? Prove it.

c. For what positive value of c would the vertex of the parabola $y = x^2 - c$ form an equilateral triangle with its roots?

22. For the graph of the quadratic $y = x^2$, investigate the area between the curve, the x-axis, and the line $x = 10$ by interactively drawing rectangles at each integer value that touch both the

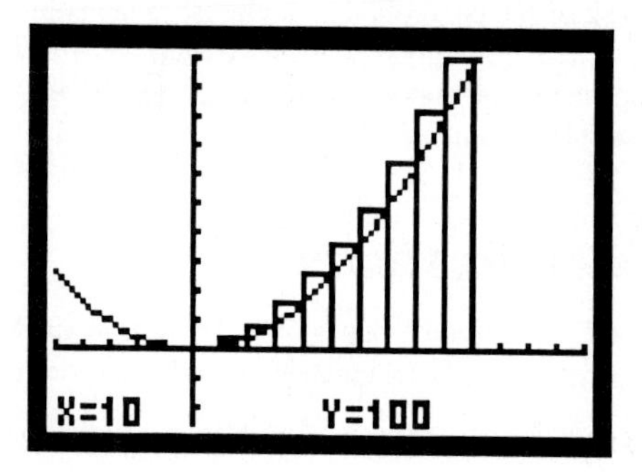

FIGURE 2.87

x-axis and the curve. See Fig. 2.87. Does it make a difference if the rectangles touch the curve on the left or the right side of the rectangle? How could you improve on this estimate of the area under the curve? Try it for $y = x^2 + h$, $h = 1$ or other values so the vertex is not at the origin.

23. To approximate the slope, m, of a tangent line drawn to a curve at a given point x, use the formula $m \approx \dfrac{f(x+h) - f(x)}{h}$ for very small values of h (i.e. .001 or .0001). Try this for the quadratic function $y = x^2$ at the point $x = 0$. What is the actual slope of any tangent line drawn through the vertex of a quadratic function? To calculate the equation of the tangent line, use the point-slope form for the equation of a line. Graph the line with the function $y = x^2$. Try this for the tangent at $x = 1$, $x = 2$, or other values. Will this work in the same way for any quadratic in the form $y = ax^2$, $y = x^2 + k$, or $y = (x - h)^2$? Try it and explain the results.

24. A triangle has 0 diagonals, a rectangle has 2 diagonals, and a pentagon has 5 diagonals. Investigate this pattern, and develop a function that gives the number of diagonals as a function of the number of sides of a polygon. Graph this function in a friendly integer window, and explore the graph using the TRACE cursor. What values make sense for this function?

2.4 Exploring More Families of Functions

In Section 2.2 we investigated **families of quadratic functions** of the form

$$y = a(x - h)^2 + k$$

compared with the **parent function**, $y = x^2$. The value of a represented the vertical stretch and flip over the x-axis, h represented the horizontal shift, and k the vertical shift. Using this form, we could predict the shape and location of the graph of any

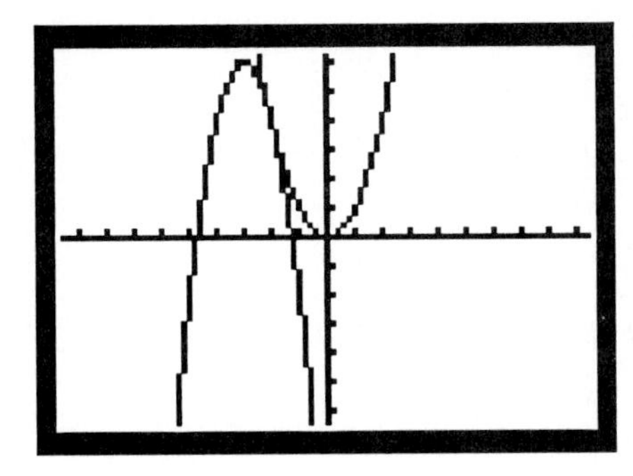

FIGURE 2.88

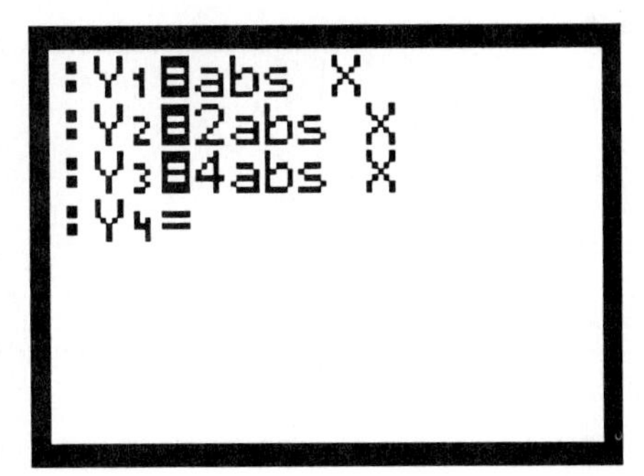

FIGURE 2.89

quadratic function. For example, the graph of $y = -2(x + 3)^2 + 6$ would be the graph of $y = x^2$ stretched vertically by a factor of 2, flipped over the x-axis, shifted 3 units to the left horizontally, and shifted 6 units up vertically. Figure 2.88 shows the confirming graph in the window [-9.6, 9.4] by [-6.4, 6.2].

In this section, we will investigate other families of functions of the general form

$$y = af(x - h) + k,$$

where several different parent functions are used to generate the family. The variables a, h, and k effect the graphs of other parent functions in the same way they effect the parent function $y = x^2$. Once you understand how these transformations take place, you can predict the shape and position of the graphs of a wide variety of functions. Often you will know what a graph should look like before you graph it on your TI-81.

EXPLORATION 1: Investigate the family of functions generated by the parent function $y = |x|$ in the form $y = a|x - h| + k$. Predict the shape and location of the graph of

$$y = -2.5\,|x + 3.2| + 1.5.$$

Compare the graphs of $y = |x|$, $y = 2|x|$, and $y = 4|x|$ on the graphics screen at the same time. (Set your calculator in **Sequence** mode so the graphs will draw one at a time.) The absolute value function is designated by the ABS key on the TI-81. Figure 2.89 shows the function entered on the Y= menu, and Fig. 2.90 shows the resulting graphs in the viewing window [-9.6, 9.4] by [-6.4, 6.2].

Use the TRACE cursor to confirm that the graph of $y = 2|x|$ is numerically the same as the graph of $y = |x|$ stretched vertically by a factor of 2, and the graph of $y = 4|x|$ is the same as the graph of $y = |x|$ stretched vertically by a factor of 4.

Figure 2.91 shows the graphs of $y = |x|$ and $y = -2|x|$, confirming that negative values for the coefficient a flip the graph over the x-axis. Figure 2.92 shows the graphs

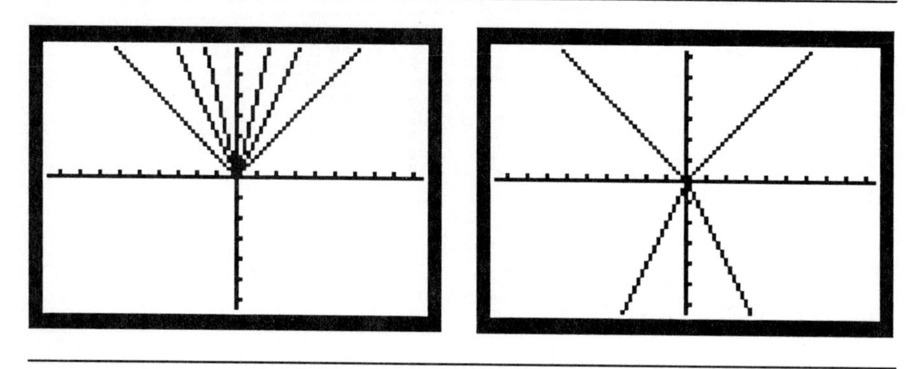

FIGURE 2.90 FIGURE 2.91

of $y = |x|$ and $y = .5|x|$, confirming that fractional values for the coefficient a represent a vertical compression of the graph.

Figure 2.93 shows the graphs of $y = |x|$, $y = |x - 3|$, $y = |x + 2|$, and $y = |x - 1|$. In the form $y = |x - h|$, the value for the variable h represents the horizontal shift of the graph and the x-coordinate of the vertex of the absolute value function.

Figure 2.94 shows the graphs of $y = |x|$, $y = |x| + 2$, $y = |x| - 3$, and $y = |x| - 6$. In the form $y = |x| + k$,, the value for the variable k represents the vertical shift of the graph and the y-coordinate of the vertex of the absolute value function.

Based on this analysis, the graph of the function $y = -2.5|x + 3.2| + 1.5$ will be the graph of $y = |x|$ stretched by a factor of 2.5, flipped over the x-axis, shifted horizontally 3.2 units to the left, and shifted vertically 1.5 units up. The vertex of the graph is at the point (-3.2, 1.5). Figure 2.95 shows the graph of this function with the graph of the parent function $y = |x|$, confirming the visual estimate. ◊

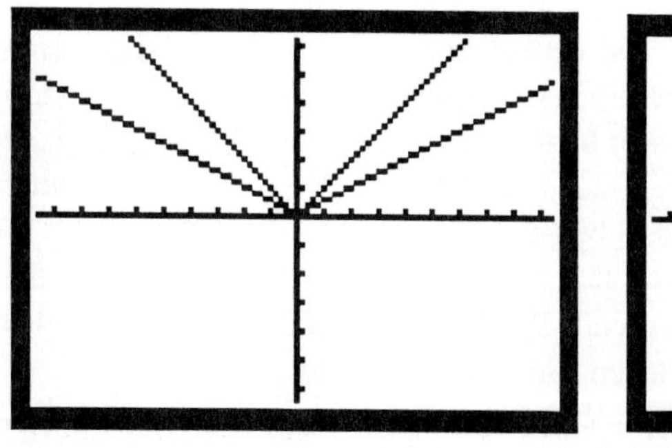

FIGURE 2.92

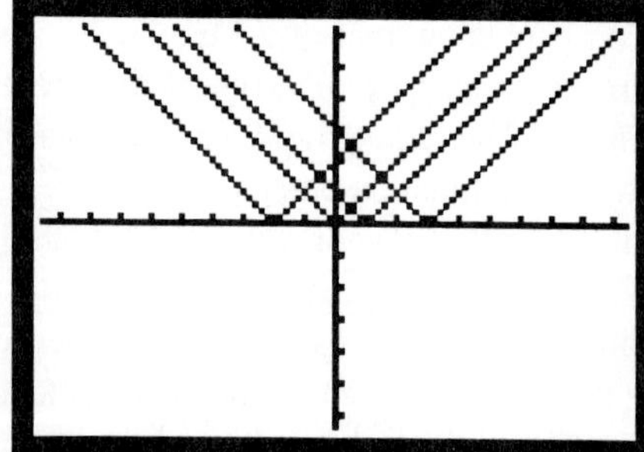

FIGURE 2.93

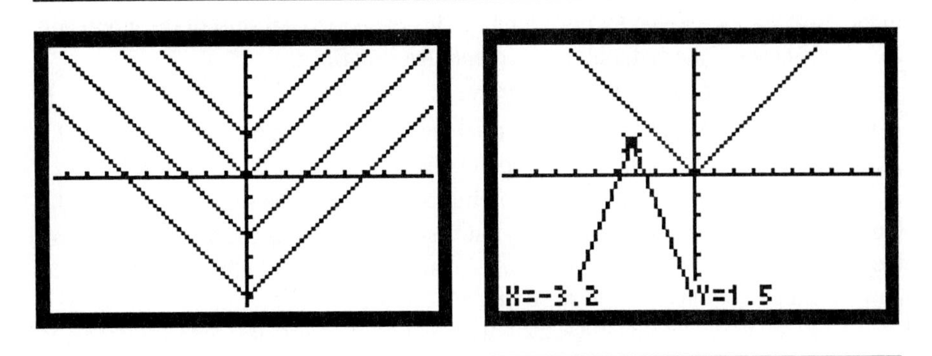

FIGURE 2.94

FIGURE 2.95

EXPLORATION 2: Investigate the family of functions generated by the parent function $y = \sqrt{x}$. Predict the shape and location of the graph of $y = \sqrt{4x+8} - 2$.

Figure 2.96 shows the graphs of $y = \sqrt{x}$, $y = 2\sqrt{x}$, $y = .5\sqrt{x}$, and $y = -3\sqrt{x}$, in the viewing rectangle [-9.6, 9.4] by [-6.4, 6.2]. (This same viewing rectangle is used for all the graphs in this exploration.) In the form $y = a\sqrt{x}$, the coefficient a affects the shape of the graph of $y = \sqrt{x}$ in the same way as with other families, a vertical stretch or compression together with a flip over the x-axis.

Notice that the graphs of these functions do not appear to the left of the y-axis. Pick an x value in this area, say $x = -2$. What is the value of $\sqrt{-2}$? Try this on your calculator. What result do you get? Why? The graph of each of these functions "starts" at a certain point. In these cases, the point is (0, 0). Use the TRACE cursor to investigate the x values that are defined for these functions and the x values that are not defined. How does your TI-81 signify that a value is not defined for a given x value?

Figure 2.97 shows the graphs of $y = \sqrt{x}$, $y = \sqrt{x-3}$, $y = \sqrt{x+4}$, and $y = \sqrt{x+8}$.

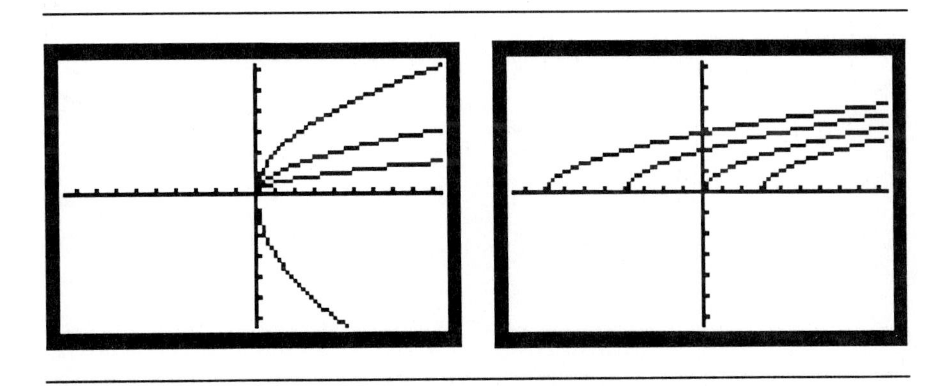

FIGURE 2.96

FIGURE 2.97

In the form $y = \sqrt{x-h}$, the variable h represents the horizontal shift of the graph and the x-coordinate of the endpoint of the graph that is defined.

Figure 2.98 shows the graphs of $y = \sqrt{x}$, $y = \sqrt{x} - 5$, $y = \sqrt{x} + 3$, and $y = \sqrt{x} + 1$. In the form $y = \sqrt{x} + k$, the variable k represents the vertical shift of the graph and the y-coordinate of the endpoint of the graph that is defined.

The function $y = \sqrt{4x+8} - 2$ is in a different form than the standard form for the families of functions, $y = af(x - h) + k$. On first glance, you might be tempted to say the graph will be stretched by a factor of 4, shifted by 8 units to the left, and shifted down 2 units. Try graphing and see if you are correct.

In several previous examples, the coefficient of the x term inside the parentheses was 1. To get this function into the correct form, we must factor out the coefficient of the x term inside the radical symbol:

$$y = \sqrt{4x+8} - 2$$
$$y = \sqrt{4(x+2)} - 2$$
$$y = \sqrt{4}\,\sqrt{x+2} - 2$$
$$y = 2\sqrt{x+2} - 2$$

The result shows the function in the form $y = af(x - h) + k$. The graph of this function will be the graph of the parent function, $y = \sqrt{x}$, stretched by a factor of 2, shifted left 2 units, and shifted down 2 units. The coordinates of the endpoint of this graph are (-2, -2). Figure 2.99 shows the resulting graphs. Move the TRACE cursor to the endpoint to confirm the coordinates of this point. ◊

EXPLORATION 3: Estimate the location and shape of the graph of the function $y = \dfrac{-3}{x-1} - 2$. What is the interpretation of the point (1, -2) for this function?

This function is a member of the family with the parent function $y = \dfrac{1}{x}$. Figure 2.100

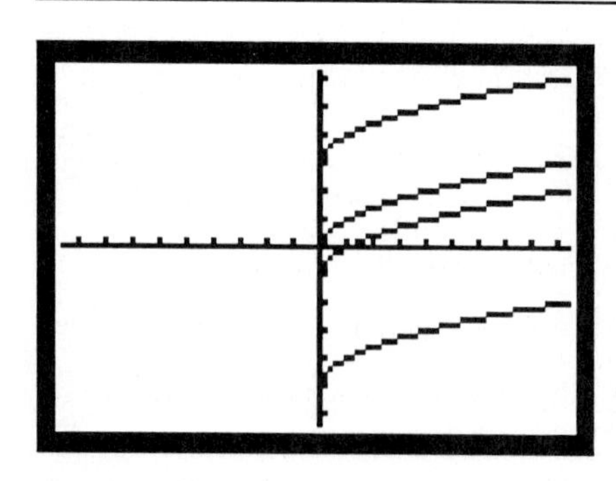

FIGURE 2.98

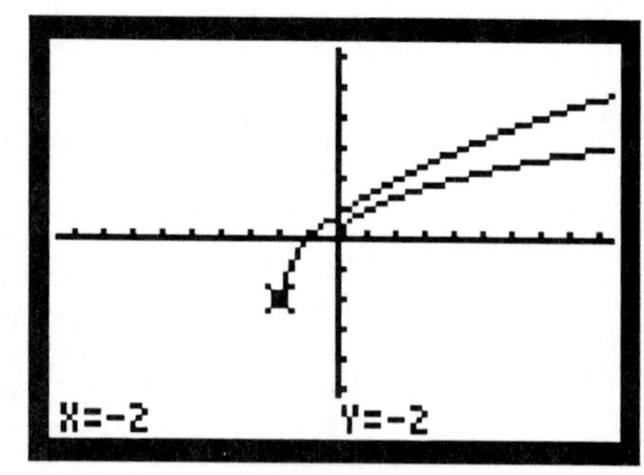

FIGURE 2.99

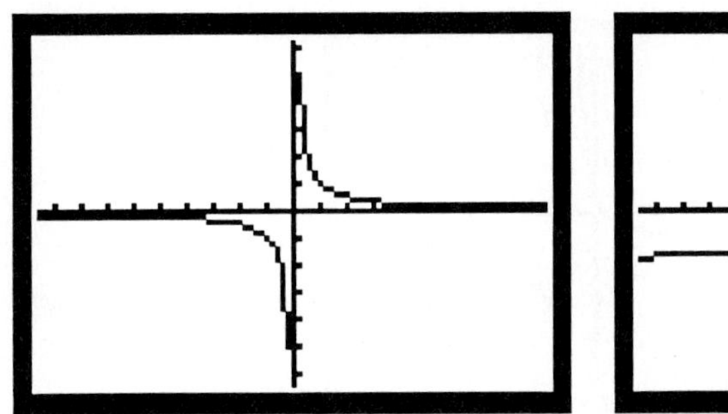

FIGURE 2.100

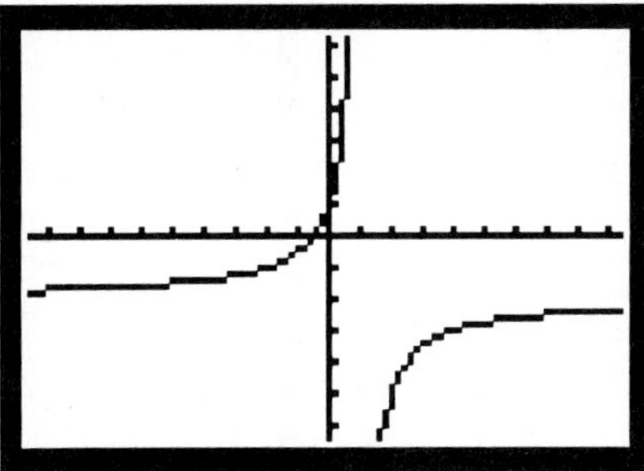

FIGURE 2.101

shows a graph of this parent function. Functions containing a fraction with the variable, x, in the denominator are called **rational functions.**

By factoring, we can change the function into the form $y = af(x - h) + k$:

$$y = \frac{-3}{x-1} - 2$$

$$y = -3\frac{1}{x-1} - 2$$

The graph of $y = \frac{-3}{x-1} - 2$ will be the same as the graph of the parent function $y = \frac{1}{x}$ stretched by a factor of 3, flipped over the x-axis, shifted 2 units to the right, and shifted 2 units down. Figure 2.101 shows the graph of this function in the friendly viewing window [-9.6, 9.4] by [-6.4, 6.2].

Undefined Values

When graphing rational functions, you must consider **if** the denominator of the fraction ever becomes zero, since division by zero is **undefined**. Move the TRACE cursor to the point $x = 1$ on the function $y = \frac{-3}{x-1} - 2$. What is the y value at this point? As the TRACE cursor passes through the values near $x = 1$, what are the y values of the function doing? Notice that the graph of the function has two disconnected branches.

Caution: Figure 2.102 shows the graph of the function $y = \frac{-3}{x-1} - 2$ in the **standard** viewing window, [-10, 10] by [-10, 10]. The vertical line connecting the two branches of this **discontinuous** function is **not** part of the graph. When $x = 1$, the denominator of the function equals zero. The x value that results in **division by zero** is called the **point discontinuity.** The vertical line you see on the graph in Fig.

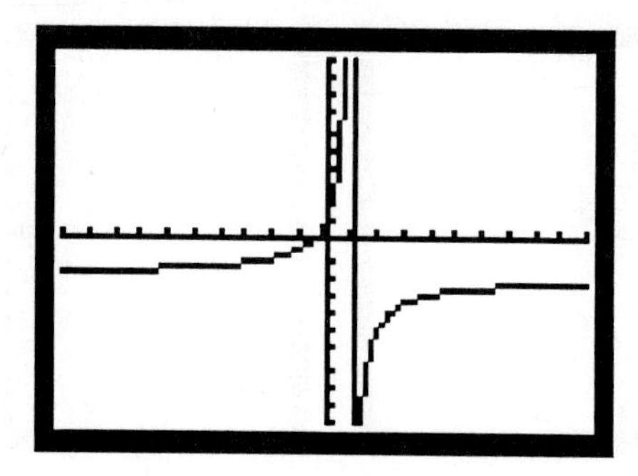

FIGURE 2.102

2.102 is a result of the scale of the viewing window, not the function. If you move the TRACE cursor across the function in the **standard** viewing window, you will see that the **exact** value $x = 1$ is not one of the values in the set of points plotted.

Sometimes a vertical and/or horizontal line is added to a rational function at a point of discontinuity. This line is called an **asymptote**. It represents a value that the function approaches, but never reaches. For the function $y = \frac{-3}{x-1} - 2$, there is a vertical asymptote at the value $x = 1$ and a horizontal asymptote at the value $y = -2$. The center of these two asymptotes is at the point (1, -2), corresponding to the point (h, k) from the general form of the function, $y = a\frac{1}{x-h} + k$. These asymptotes have been added to the graph of $y = \frac{-3}{x-1} - 2$ (in the viewing window [-9.6, 9.4] by [-6.4, 6.2]) in Fig.

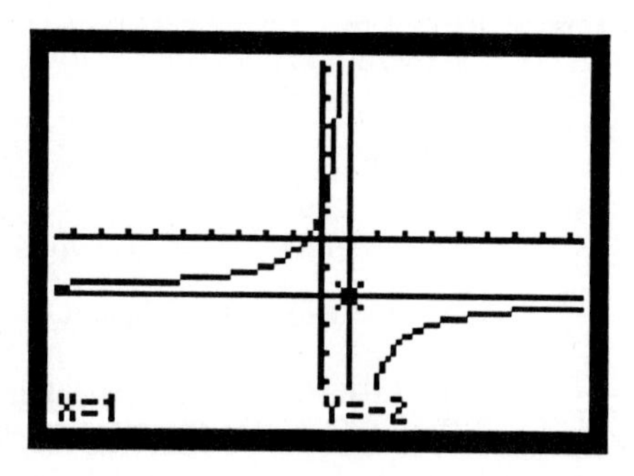

FIGURE 2.103

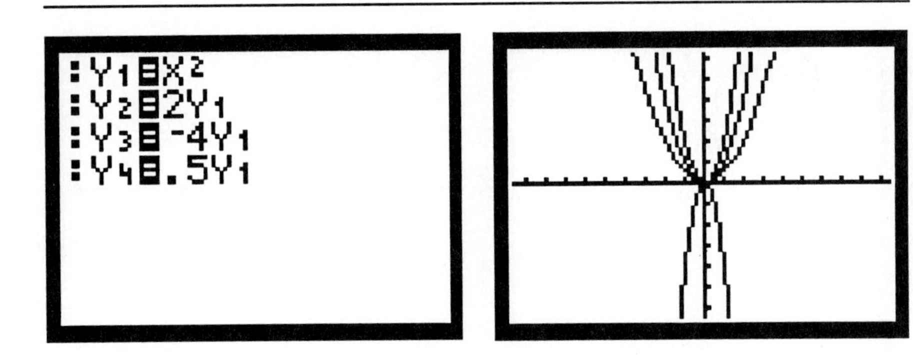

FIGURE 2.104

FIGURE 2.105

2.103 by graphing the line $\mathbf{Y_2}$ = **-2** and using the **Line** command. Can you draw the vertical asymptote as a function? ◊

ANOTHER APPROACH TO FAMILIES OF FUNCTIONS

The general form for translations of functions leads to another approach to vertical stretching and shifting of families of functions. The general form is

$$y = af(x - h) + k.$$

In this form, the parent function, f, can be represented on the [Y=]) menu by $\mathbf{Y_1}$. Figure 2.104 shows the [Y=] menu for the $y = x^2$ family using different values for the variable a. Figure 2.105 shows the resulting graphs in the viewing window [-9.6, 9.4] by [-6.4, 6.2].

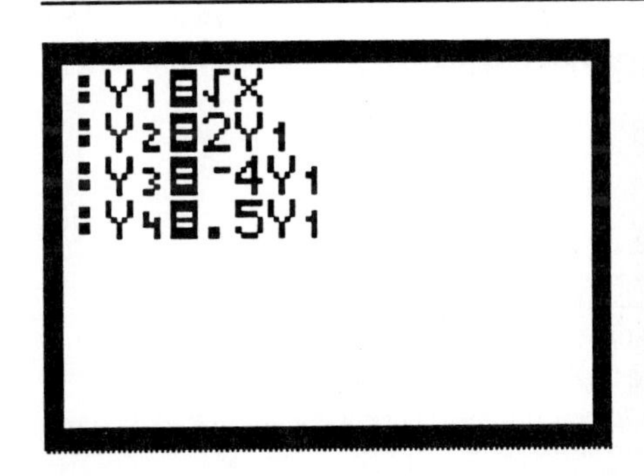

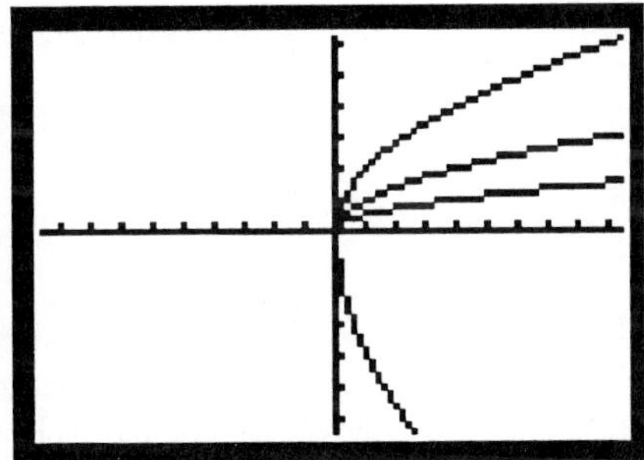

FIGURE 2.106

FIGURE 2.107

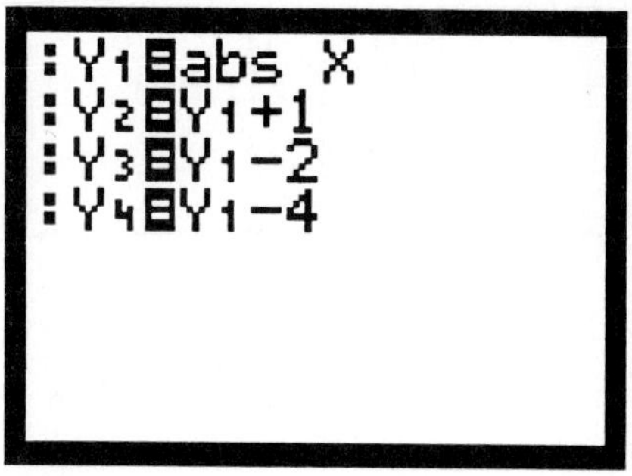

```
:Y1=abs X
:Y2=Y1+1
:Y3=Y1-2
:Y4=Y1-4
```

FIGURE 2.108

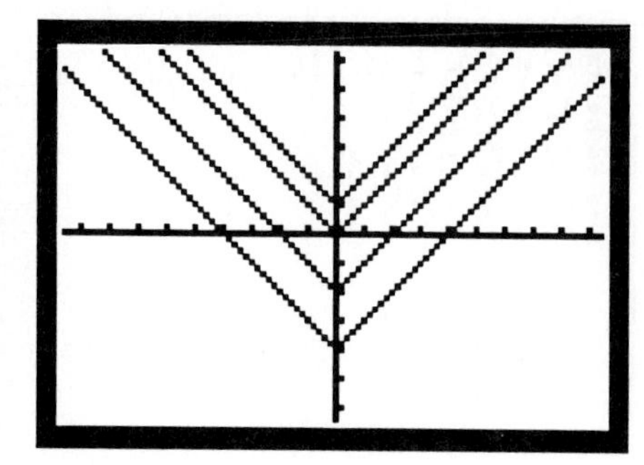

FIGURE 2.109

By changing the parent function, for example to $y = \sqrt{x}$, the new set of graphs under the same vertical stretching factors is produced (see Figs. 2.106 and 2.107). This technique will also work for vertical translations (see Figs. 2.108 and 2.109.)

Can you represent a horizontal shift of the function in the same way? Explain.

Problems

1. Use the parent function $y = |x|$ to describe the shape and location of the graphs of the following without the graphing calculator, and then confirm your estimate using one. List the coordinates of the vertex of each function, and confirm this using the [TRACE] cursor.

   a. $y = |x| - 5$ **b.** $y = |x - 5|$ **c.** $y = 2|x|$

   d. $y = -|x|$ **e.** $y = 3|x + 3.2| - 8.7$ **f.** $y = -2|3x + 6| + 1.5$

2. Use the parent function $y = \dfrac{1}{x}$ to describe the shape and location of the graphs of the following without the graphing calculator, and then confirm your estimate using one. List the coordinates of the intersection and the equations of the horizontal and vertical asymptotes for each function.

   a. $y = \dfrac{3}{x}$ **b.** $y = \dfrac{-1}{x}$ **c.** $y = \dfrac{1}{x} + 3$

   d. $y = \dfrac{2}{x - 4}$ **e.** $y = -5 + \dfrac{2}{3 - x} - 4$ **f.** $y = \dfrac{3}{3x - 6} - 4$

3. Use the parent function $y = \sqrt{x}$ to describe the shape and location of the graphs of the following without the graphing calculator, and then confirm your estimate using one. List the coordinates of the endpoint of the graph, and confirm this value using the [TRACE] cursor.

   a. $y = 2\sqrt{x}$ **b.** $y = \sqrt{x - 2}$ **c.** $y = \sqrt{x} - 2$

d. $y = -\sqrt{x}$ **e.** $y = -4\sqrt{3-x} - 2$ **f.** $y = \sqrt{9x-36} + 2.5$

4. Explore the graph of the general function $y = |x - a| + |x - b|$ for the following. Determine the effect of a and b on the shape and position of the graph, and make a general rule for any values. What if $a = b$?

a. $y = |x - 1| + |x + 1|$ **b.** $y = |x + 2| + |x - 1|$

c. $y = |x - 2| + |x - 1|$ **d.** $y = |x - 3| + |x - 1|$

5. Use the definition of absolute value to change $|x| + |y| = 1$ so it can be entered in the [Y=] menu and graphed. Predict the shape and location of the graphs of the following functions, and check by graphing them.

a. $|x + 2| + |y| = 1$ **b.** $|x - 2| + |y| = 1$

c. $|x| + |y - 1| = 1$ **d.** $|x - 1| + |y - 1| = 1$

6. Investigate the following curves and explain their behavior arithmetically.

a. $y = \frac{1}{x^2}$ **b.** $y = \frac{1}{|x|}$ **c.** $y = \frac{1}{\sqrt{x}}$ **d.** $y = \frac{|x|}{x}$

7. Check the following algebra identities graphically. Confirm that they are identities by using the [TRACE] cursor on several points to show the two graphs coincide (are on top of each other). Confirm the identities algebraically by transforming one side of the equation into the other.

a. $\frac{1}{x} + \frac{1}{x+1} = \frac{2x-1}{x(x-1)}$ **b.** $\sqrt{2x}\,\sqrt{x-1} = \sqrt{2x^2 - 2x}$

c. $\frac{\sqrt{x}}{\sqrt{x-1}} = \sqrt{\frac{x}{x-1}}$

8. Solve the following graphically, and check your solution algebraically.

a. $3x = \frac{9}{x} + 26$ **b.** $\sqrt{x^2 - 3x} = 2$

c. $\frac{4}{x-1} = \frac{5}{2x-2} + \frac{3x}{4}$

9. Use your knowledge of transformations to find the inequality that is **NOT** true for all x. Confirm your analysis by graphing both sides of the inequalities.

a. $|x + 3| \le |x| + 3$

b. $|x - 3| \le |x| - 3$

c. $|x + 2 + 3| \le |x| + |2| + |3|$

10. Graph the function $y = \sqrt{x}$ and plot the point (4, 0). Estimate the point on the curve that is closest to the given point. Some zooming in may be necessary to get a good estimate. (*Hint*: Using the **Line** command interactively may help.)

11. Graph the function $y = \frac{1}{x}$ and plot the point (2, 2). Estimate the point on the curve that is closest to the given point. Some zooming in may be necessary to get a good estimate. Try to transform $y = \frac{1}{x}$ so it passes through the point (2, 2).

12. The equation $x^2 = \sqrt{x}$ has two solutions ($x = 0$ and $x = 1$; check by graphing). By graphing and from our understanding of transformations, we can see that $y = x^2 + 1$ and $y = \sqrt{x}$ have no solution (i.e. intersection points). Investigate when the graphs of the functions $y = ax^2 + 1$ and $y = a\sqrt{x}$ have only one solution. That is, for what value of a will the graphs of these curves only touch at one point? Find the value of a and the coordinates of the point of intersection. Change the original problem to investigate when $ax^2 + 1 = \sqrt{ax}$ has only one solution. Explain why these answers differ.

13. The graphs of the functions $y = -x^2$ and $y = \frac{1}{x}$ do not intersect in the first quadrant (i.e. for $x \geq 0$). Experiment with the graphs of $y = -ax^2 + a$ and $y = \frac{1}{ax} - a$ to find the value of a that makes the curves touch just once in the first quadrant. Give the value of a and the coordinates of the point of intersection.

14. Explore the graphs of the general form of the function $y = a|b|x|| - c$. Determine the effect of the variables a, b, and c on the shape and position of the graph. Experiment with some functions of your own so you can predict the shape and location of the figures formed. What if the value for c is negative? Try extending the pattern to include more imbedded absolute value operations. What happens?

a. $y = 1|1|x|| - 1$ **b.** $y = 2|3|x|| - 1$

c. $y = 3|2|x|| - 1$ **d.** $y = 1|1|x|| - 2$

15. Try to predict the graph of $y = |x| + \sqrt{1 - x^2}$ and $y = |x| - \sqrt{1 - x^2}$. Can you draw an arrow through it? Can you shade it?

FOR ADVANCED ALGEBRA STUDENTS

16. Graph the function y= Int(x) in **Dot** mode. (The **greatest integer function, 4: Int,** is found on the MATH key under the **NUM** submenu.) This is called the greatest integer function because the function value is the largest integer less than or equal to the value of x. Figure 2.110 shows a number-line model for the greatest integer function. The function value is the integer just to the left of the x value on the number line. Use the parent function y = Int(x) to determine the shape and location of the graphs of the following functions without the graphing calculator, and then check using one.

a. y = Int(x) – 3 **b.** y = Int(x + 2) **c.** y = 3 Int(x) **d.** y = Int(3x)

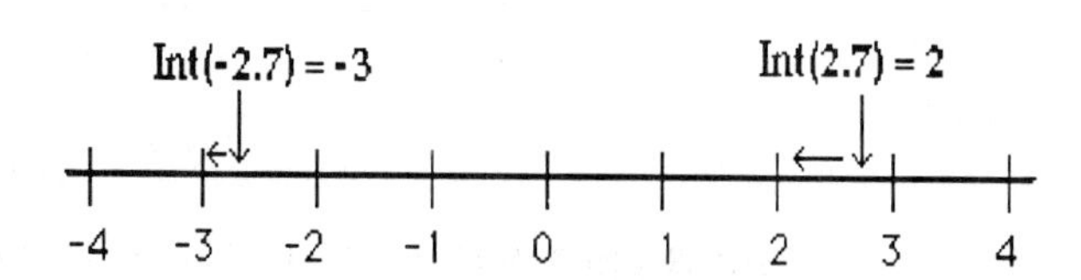

FIGURE 2.110

17. Use the parent function $y = \sin x$ to graph the following without the graphing calculator, and then check using one.

a. $y = 3\sin x$ **b.** $y = \sin 3x$ **c.** $y = \sin(x - 3)$ **d.** $y = \sin x + 3$

18. Graph the following inequalities first in a single variable (e.g. $\mathbf{Y_1}$ **= abs (X – 4)5)** and then in two variables (e.g. $\mathbf{Y_1}$ **= abs (X – 4)** and $\mathbf{Y_2}$ **= 5**). Confirm the results algebraically and explain the answer.

a. $|x - 4| < 5$ **b.** $|x + 6| > 3$ **c.** $|3x - 5| < 2$ **d.** $|1 - 2x| > 9$

19. Solve the inequality $|x + 3| < |x|$ graphically either in a single variable or in two variables, and explain your answer using the definition of absolute value.

20. Find the solutions of the equation $|x - 5|^2 + 2|x - 5| - 3 = 0$ graphically, and confirm your results algebraically.

21. Investigate the following functions graphically. These functions are given in the quadratic form $y = ax^2 + bx + c$. Determine if any of these functions generate "families" when you experiment with different values for the coefficients a, b, and c. (*Hint*: Change one coefficient at a time to determine its effect.)

a. $y = a|x|^2 + b|x| + c$ **b.** $y = \left(\frac{a}{x}\right)^2 + \frac{b}{x} + c$

c. $y = a\sqrt{x^2} + b\sqrt{x} + c$ **d.** $y = a(\sin x)^2 + b \sin x + c$

22. Investigate the following functions graphically to see if they have any "families." What effect do a and b have on the shape and position of the graph? Can you predict any turning points? What happens if the value of b is negative? What determines if the graph is symmetric around the y-axis or the origin? (*Hint*: First let $a = 1$ and change b to assist your investigation; then try different values for a.)

a. $y = ax^2 + \frac{b}{x^2}$ **b.** $y = ax + \frac{b}{x}$

c. $y = a\sqrt{x} + \frac{b}{\sqrt{x}}$ **d.** $y = a|x| + \frac{b}{|x|}$

23. Graph the following equations to see if they represent identities. Explain any similarities or differences algebraically.

a. $||x|| = |x|^2$ **b.** $\sin(\sin x) = (\sin x)^2$

c. $\sqrt{x^2} = \sqrt{x}\,\sqrt{x}$ **d.** $|\sin x| = (\sin x)^2$

e. $|\sqrt{x}| = \sqrt{|x|}$ **f.** $\frac{1}{|x|} = \frac{1}{x^2}$

24. The following investigation is the result of a problem posed in the *AMTNYS New York State Math Teacher's Journal* (*41*, 1, 1991), p. 63. Investigate the following function graphically: $y = a\big||x - b| + c\big| + d|x - e| + f$.

a. What happens if d and f are equal?

b. What effect do a and d have when $d = f$?

c. What happens to the slope of the "wings" and the slope of the "vee" if a and d are changed?

d. Pay particular attention to the sum and difference of a and d.

e. What effect do b, c, e, and f have?

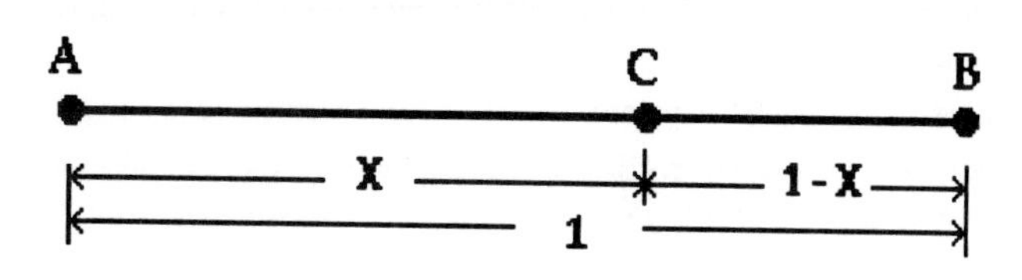

FIGURE 2.111

25. Investigate the function $y = |a - |a - |a - |x|||$ graphically for different values of a, and explain what effect this variable has on the graph. How would the graph appear if we add more absolute value operators to this function? Explain the function algebraically.

26. Investigate $|xy| = 1$ using the definition of absolute value and the idea of symmetry. Can you draw the graph of this relationship in two "pieces"?

27. When a line segment AB is divided by the point C such that $\frac{AC}{CB} = \frac{AB}{AC}$, then the length of line segment AC is called the **golden ratio** (see Fig. 2.111). If we let AB = 1, AC = x, and CB = $1 - x$, then we get the following ratio:

$$\frac{x}{1-x} = \frac{1}{x}.$$

Solve this equation graphically to find the decimal approximation of the "golden ratio," and solve it algebraically to find the exact value.

28. Glenda Starbuck is retiring after 40 years of faithful service. All her colleagues contributed equal amounts to buy a special gift that costs \$100. Before the gift was presented at the party, one more person contributed money toward the gift. This last donation reduced the amount of each equal share by \$5. How many friends were in the original group? Solve the problem graphically, and then confirm your solution algebraically.

3

Exploring Polynomial, Rational, and Radical Functions

3.1 Exploring Polynomial Functions

The quadratic functions we explored in Chapter 2 are polynomials of degree 2, and the linear functions we explored in Chapter 1 are polynomials of degree 1. In this section we will explore polynomials of higher degree. The graphs of each different degree of polynomial have certain predictable characteristics that will help us understand the shape and position of the graph in the plane.

EXPLORATION 1: Compare the graphs of $y = x$, $y = x^3$, and $y = x^5$ on the screen at the same time. What happens to the function values (y values) of the graph as the value of x approaches $\pm\infty$ (positive and negative infinity)? Make and test a conjecture about the shape and position of the graphs $y = x^7$, $y = x^9$, $y = x^{11}$, and $y = x^{13}$. Do these graphs have any points in common? Make and test a conjecture about the shape and position of the graphs of $y = -x$, $y = -x^3$, and $y = -x^5$. Test your conjecture, and make a general rule about the shape and position of the graphs of $y = x^n$ and $y = -x^n$ for n an odd positive integer.

Figure 3.1 shows the graphs of $y = x$, $y = x^3$, and $y = x^5$ in the viewing window [-4.8, 4.7] by [-3.2, 3.1]. In each case, the graphs occur in the first and third quadrants. Use the TRACE cursor to explore the values of the functions numerically. As x gets larger and larger, the y values of all three functions increase. Figure 3.2 shows the TRACE cursor scrolled to the point (9, 59,049) on the function $y = x^5$ in the

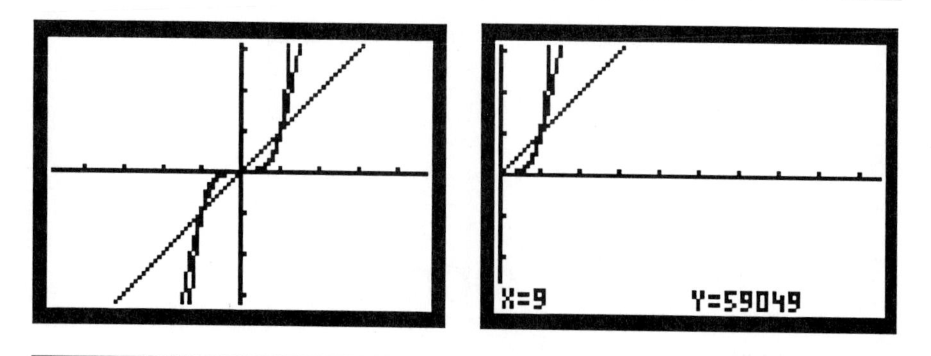

FIGURE 3.1

FIGURE 3.2

viewing window [0, 9.5] by [-3.2, 3.1] (the cursor is off the screen). The function values continue to increase for larger values of x. The same result is seen on the graph of $y = x$ and $y = x^3$. In each case, as $x \to \infty, f(x) \to \infty$.

As x becomes more negative, the values of the functions approach $-\infty$. Move the [TRACE] cursor to the point (-8, -512) on the graph of $y = x^3$. Switch to the graph of $y = x^5$; the coordinates of the point are (-8, -32,768).

Change the [RANGE] of the viewing window to [-100, 100] by [-10,000, 10,000], and draw the graphs of $y = x$, $y = x^3$, and $y = x^5$. Activate the [TRACE] cursor to explore the function values in this viewing window as x gets larger or smaller. Figure 3.3 shows the cursor on the graph of $y = x^5$. In general, as $x \to \infty, f(x) \to \infty$ for this function. The behavior of a function as $x \to \pm\infty$ is called the **end behavior of the function.**

Figure 3.4 shows the graphs of $y = x$, $y = x^3$, and $y = x^5$ in the viewing window [-4.8, 4.7] by [-3.2, 3.1] with the [TRACE] cursor on the graph of $y = x^3$ at the point (1, 1). By switching from one function to the other using the [▲] and [▼] arrow keys,

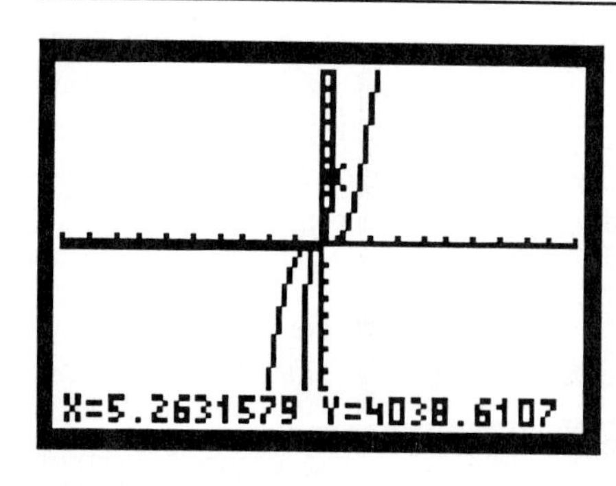

FIGURE 3.3

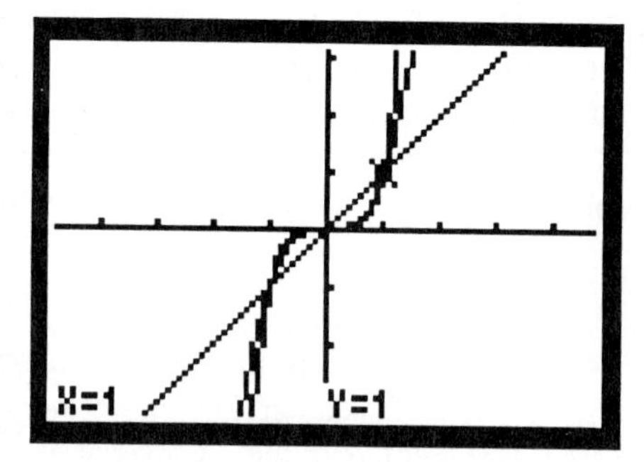

FIGURE 3.4

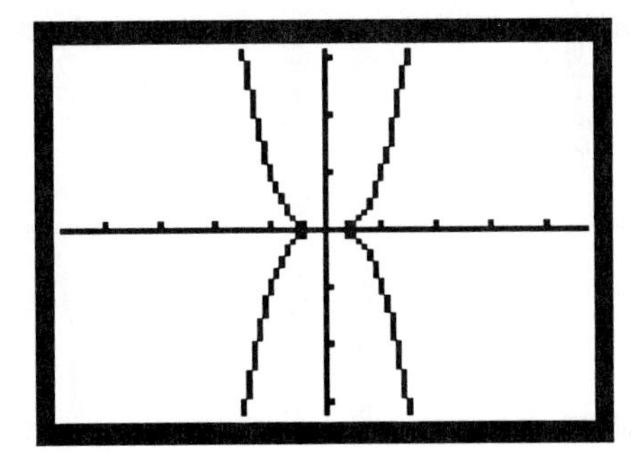

FIGURE 3.5

we can see that the point (1, 1) is common to all three graphs. Move the cursor to the points (0, 0) and (-1, -1) to show that these points are also on all three graphs. Check each of these points numerically to confirm that they are common to all the graphs.

Draw the graphs of $y = x^7$, $y = x^9$, $y = x^{11}$, and $y = x^{13}$ in the viewing window [-4.8, 4.7] by [-3.2, 3.1], and confirm that these graphs have the points (-1, 1), (0, 0), and (1, 1) in common. In general, all functions of the form $y = x^n$, for n an odd positive integer, have these three points in common. (Why?)

Figure 3.5 shows the graphs of $y = x^3$, and $y = -x^3$ in the viewing window [-4.8, 4.7] by [-3.2, 3.1]. Notice that the graph of $y = -x^3$ is the graph of $y = x^3$ reflected across the x-axis. A negative coefficient on the lead term of a higher degree polynomial has the same effect as the negative coefficient in $y = -x^2$.

Figure 3.6 shows the same effect for the pairs of functions $y = x^5$ and $y = -x^5$, and $y = x^7$, and $y = -x^7$. Using the TRACE cursor, confirm that, in general, for functions

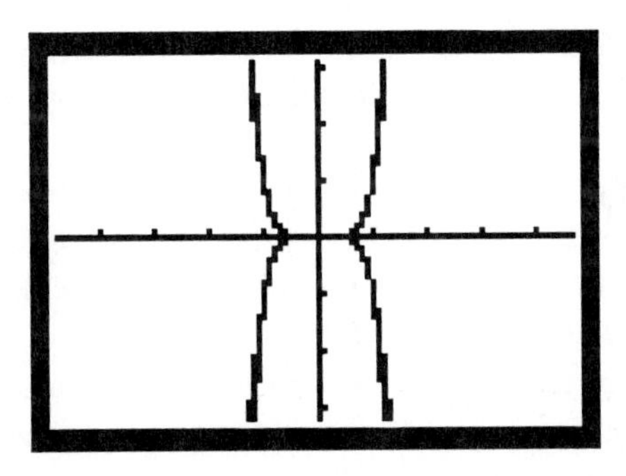

FIGURE 3.6

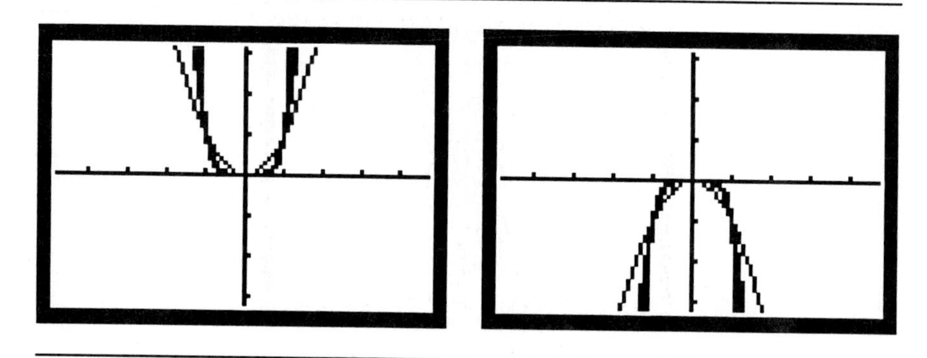

FIGURE 3.7

FIGURE 3.8

of the form $y = -x^n$, for n an odd positive integer, as $x \to \infty$, $f(x) \to -\infty$, and as $x \to -\infty$, $f(x) \to \infty$.

Test this conjecture for higher degree polynomials, like $y = -x^{15}$. Use the TRACE cursor to show that these functions have the points (-1, 1), (0, 0), and (1, -1) in common. Why? Confirm numerically that these points are common to all the functions.

◊

EXPLORATION 2: Explore the graphs of $y = x^2$, $y = x^4$, $y = x^6$, and $y = x^8$ and $y = -x^2$, $y = -x^4$, $y = -x^6$, and $y = -x^8$ to determine common behavior, common points, end behavior, and general rules for the shape and position of polynomial graphs of even positive integer degree.

Figure 3.7 shows the graphs of $y = x^2$, $y = x^4$, $y = x^6$, and $y = x^8$ in the [-4.8, 4.7] by [-3.2, 3.1] viewing window. All of the graphs appear in the first and second quadrants. The TRACE cursor confirms that the points (-1, 1), (0, 0), and (1, 1) are common to all the graphs. Confirm this numerically. Will this be true for $y = x^{200}$?

Figure 3.8 shows the graphs of $y = -x^2$, $y = -x^4$, $y = -x^6$, and $y = -x^8$ in the [-4.8, 4.7] by [-3.2, 3.1] viewing window. These graphs are the reflections of the graphs of $y = x^2$, $y = x^4$, $y = x^6$, and $y = x^8$ across the x-axis. This is the same result as before. The points (-1, -1), (0, 0), and (1, -1) are common to these graphs. Why? Confirm your results numerically.

Use the TRACE cursor to confirm that for functions of the form $y = x^n$, for n representing a positive even integer, as $x \to \pm\infty$, $f(x) \to \infty$. This means that as the absolute value of x gets larger, the function values always get larger. Why? In visual terms, this means the graphs open upward.

For functions of the form $y = -x^n$, for n a positive even integer, as $x \to \pm\infty$, $f(x) \to -\infty$. This is the reflection of the previous case. As x gets larger or smaller, the function values get smaller (more negative). Why? Visually, this means the graphs open downward.

◊

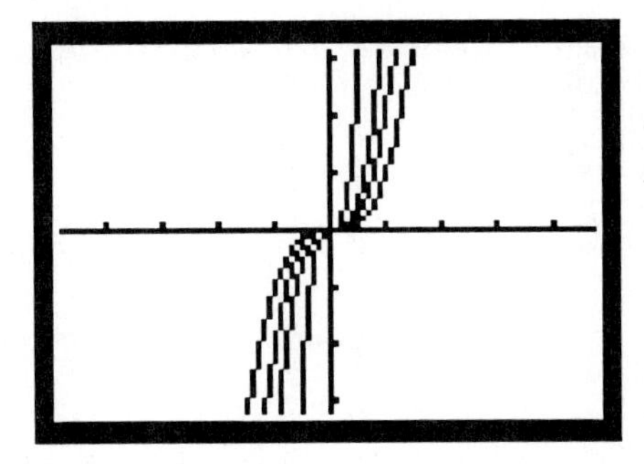

FIGURE 3.9

EXPLORATION 3: Make a conjecture about the shape and position of the graph of $y = ax^3 + k$ compared with the graph of $y = x^3$, and the graph of $y = ax^4 + k$ compared with the graph of $y = x^4$. Test your conjecture for several different values for a and k. Make a generalization based on your evidence and test it on several higher degree polynomials of the same type.

CONJECTURE: Based on work with the graph of $y = ax^2$ compared with $y = x^2$, if $|a| > 1$, the graphs of $y = ax^3$ and $y = ax^4$ will be stretched vertically and if $|a| < 1$, then the graphs of $y = ax^3$ and $y = ax^4$ will be compressed vertically.

Figure 3.9 shows the graphs of $y = x^3$, $y = 2x^3$, $y = 4x^3$, and $y = 20x^3$ in the [-4.8, 4.7] by [-3.2, 3.1] viewing rectangle. Figure 3.10 shows the graphs of $y = x^3$, $y = .5x^3$, $y = .25x^3$, and $y = .1x^3$ in the [-4.8, 4.7] by [-3.2, 3.1] viewing rectangle. Use the TRACE cursor to confirm that these graphs have been stretched and compressed

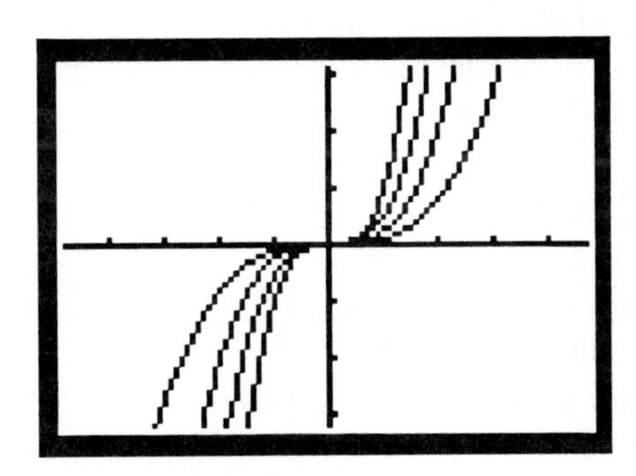

FIGURE 3.10

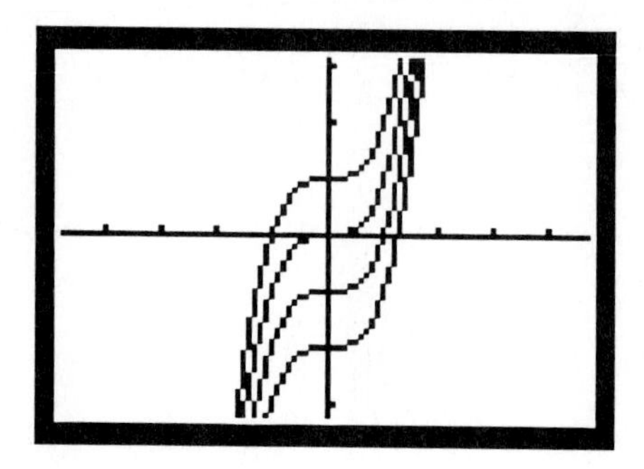

FIGURE 3.11

by the factor equal to the $|a|$. If a is negative, the graph will be reflected about the x-axis.

CONJECTURE: Based on work with the graph of $y = x^2$ compared with $y = x^2 + k$, the graphs of $y = x^3 + k$ and $y = x^4 + k$ will be shifted vertically by k units. The shape of the graph will be identical to the parent function; only the vertical position will change.

Figure 3.11 shows the graphs of $y = x^3$, $y = x^3 + 1$, $y = x^3 - 1$, and $y = x^3 - 2$, and Fig. 3.12 shows the graphs of $y = x^4$, $y = x^4 + 1$, $y = x^4 - 1$, and $y = x^4 - 2$ in the [-4.8, 4.7] by [-3.2, 3.1] viewing rectangle. In each case, adding or subtracting a constant amount from the parent function shifts the graph up or down by the amount of the constant; the shape of the graph is unchanged.

EXPLORATION 4: Make and test a conjecture about the shape and position of the

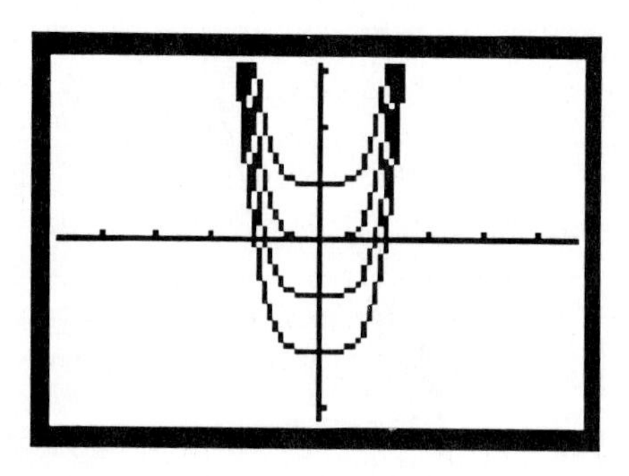

FIGURE 3.12

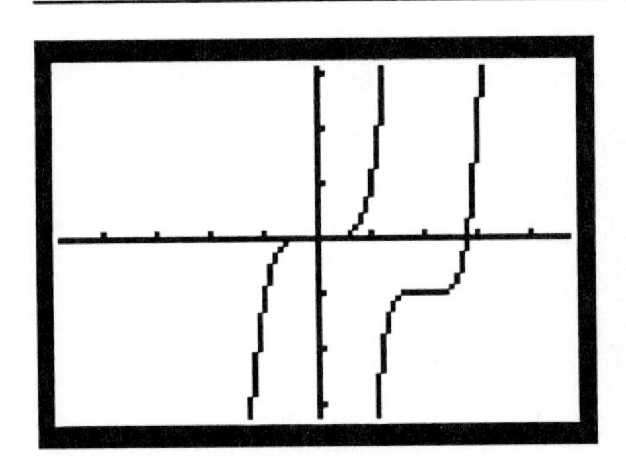

FIGURE 3.13

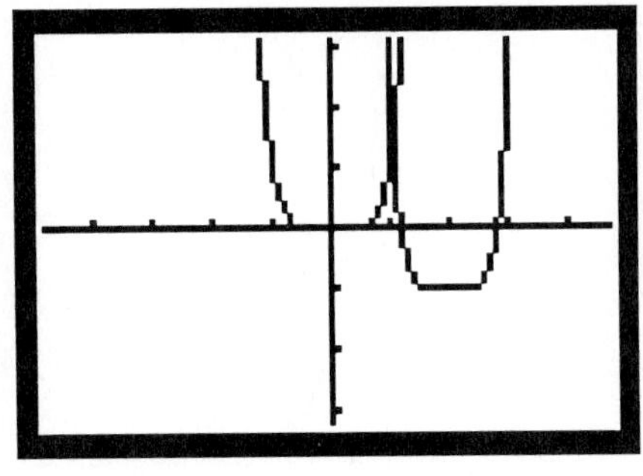

FIGURE 3.14

graph of $y = a(x - h)^5 + k$ compared with the graph of $y = x^5$ and $y = a(x - h)^6 + k$ compared with $y = x^6$.

CONJECTURE: Based on previous experiments, the graph of $y = a(x - b)^5 + k$ or $y = a(x - b)^6 + k$ should be the graph of $y = x^5$ or $y = x^6$, respectively, stretched by a factor of a, shifted h units horizontally, and shifted k units vertically.

Figure 3.13 shows the graphs of $y = x^5$ and $y = 3(x - 2)^5 - 1$ in the [-4.8, 4.7] by [-3.2, 3.1] viewing rectangle. Figure 3.14 shows the graphs of $y = x^6$ and $y = 3(x - 2)^6 - 1$ in the same [-4.8, 4.7] by [-3.2, 3.1] viewing rectangle. In both cases, the parent function is stretched by a factor of 3, shifted 2 units to the right, and shifted 1 unit down. These dilations and translations follow the same pattern seen previously with quadratic equations.

REAL ROOTS OF POLYNOMIALS

Linear functions of the form $y = ax + b$ $(a \neq 0)$ have one x-intercept. A general function f(x) may have no x-intercept, one x-intercept, or any other number of x-intercepts. These points, where a function crosses the x-axis, are called the **roots** or **zeros of the function.** In general, the coordinates of the roots of a function are (x, 0), hence the name "zero of the function." Another way to view a root is as the solution to the equation $f(x) = 0$, for any function $f(x)$. Any linear function with slope other than zero (i.e. $a \neq 0$) will have **exactly** one root that is a real number, called a **real root.** Why? Is there any way a linear function could have more than one real root?

Quadratic functions of the form $y = ax^2 + bx + c$ can have more than one real root.

EXPLORATION 5: Explore the number of real roots for the following quadratic functions:

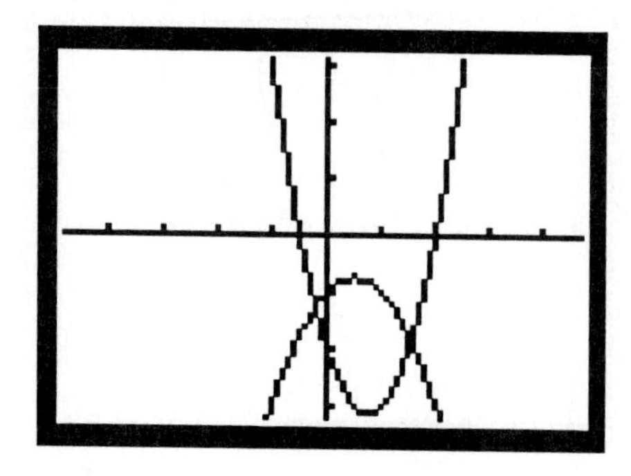

FIGURE 3.15

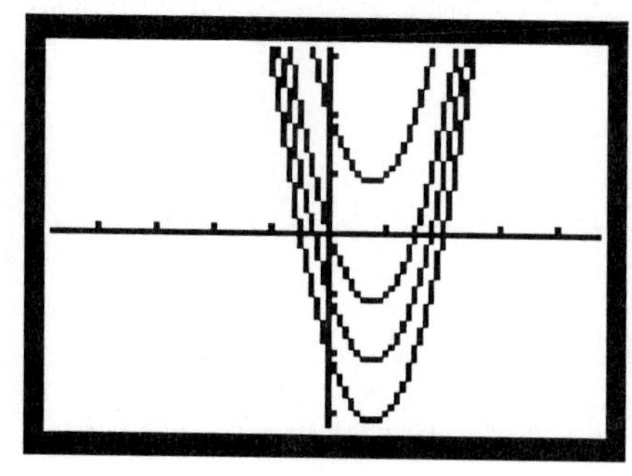

FIGURE 3.16

$$y = 2x^2 - 3x - 2, \qquad y = -x^2 + x - 1, \qquad y = x^2 + 2x + 1.$$

Conclude your investigation with a general statement about the number of real roots a quadratic function can have.

Figure 3.15 shows the graphs of $y = 2x^2 - 3x - 2$ and $y = -x^2 + x - 1$ in the [-4.8, 4.7] by [-3.2, 3.1] viewing window. The function $y = 2x^2 - 3x - 2$ has two real roots at $x = -.5$ and $x = 2$. (Confirm this with the TRACE cursor.) The function $y = -x^2 + x - 1$ has no real roots (it does not cross the x-axis).

Imagine translating either of these functions vertically by changing the constant term. Figure 3.16 shows the original function $y = 2x^2 - 3x - 2$ and three variations created by changing the constant term to -1, 0, and +2. The original function and three variations are

$$y = 2x^2 - 3x - 2,$$
$$y = 2x^2 - 3x - 1,$$
$$y = 2x^2 - 3x + 0,$$
$$y = 2x^2 - 3x + 2.$$

What happens to the number of real roots as the function is shifted upward? For each of the first three functions, there are two real roots. For the last function, $y = 2x^2 - 3x + 2$, there are no real roots. Could there be only one real root?

Figure 3.17 shows the graph of $y = x^2 + 2x + 1$ in the [-4.8, 4.7] by [-3.2, 3.1] viewing rectangle. The TRACE cursor seems to indicate that there is only one real root at the point (-1, 0).

If we factor the expression $0 = x^2 + 2x + 1$, we get $0 = (x + 1)^2$, which is the same as the expression $0 = (x + 1)(x + 1)$. By setting each of these factors equal to zero, we can see that there are really two roots, $x = -1$ and $x = -1$, which are identical. In this case, we say that this function has a single root **of multiplicity two.** The same situation can occur with higher degree polynomials. Figure 3.18 shows the graphs of $y = (x - 1)^3$ and $y = (x + 2)^6$, which have multiple roots at $x = 1$ and $x = -2$, respectively.

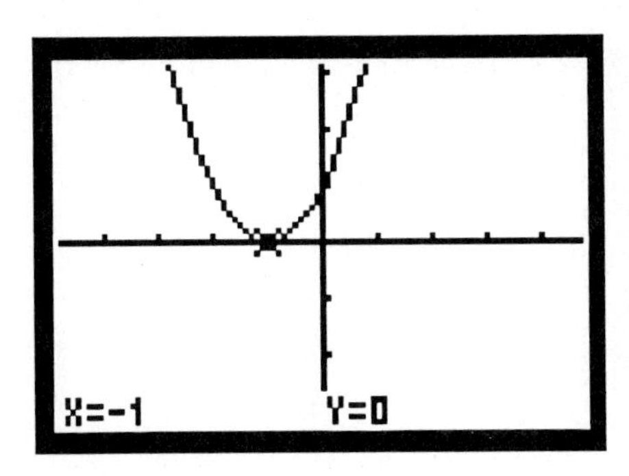

FIGURE 3.17

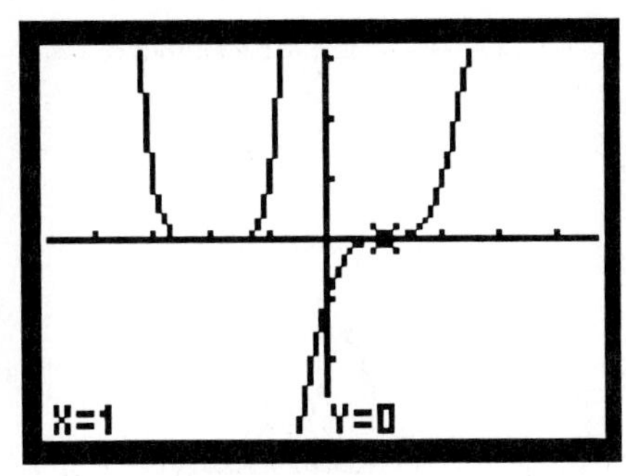

FIGURE 3.18

GENERAL STATEMENT: A quadratic function (a polynomial of degree 2) can have zero or two real roots. If the quadratic function just touches the axis, then it has two identical roots and can be factored to the form $y = a(x - h)^2$. ◊

It appears that linear functions (polynomials of degree 1) can have one real root and quadratic functions (polynomials of degree 2) can have as many as two real roots. What will happen for polynomials of degree 3, 4, 5, or higher?

EXPLORATION 6: Explore the shape and number of real roots for the following third degree (cubic) functions as compared with the parent cubic function $y = x^3$:

$$y = x^3 + 3x^2 - 2x - 5, \qquad y = -2x^3 - 5x^2 - x + 2, \qquad y = .5x^3 + x + 2.$$

Conclude your investigation with a general statement about the number of real roots a cubic function can have, when the function is **increasing** or **decreasing,** and the number of **turning points** (**local maximum** and/or **local minimum** points) on the curve.

A function is said to be **increasing** on an interval if the function values are getting larger as the x values increase. Visually, this means that as the graph of the function is drawn from left to right, the vertical motion of the graph is upward. A function is **decreasing** on an interval if the function values are getting smaller as the x values increase. As the graph of the function is drawn, the vertical motion is downward.

A **local maximum** or **local minimum** is a place on the curve where you see a **turning point.** That is, these are points where the vertical motion (increasing or decreasing) of the graph reverses direction. These local maximum and minimum points are not the **absolute** highest or lowest point on the graph, as with absolute maximum or minimum points, but only the highest or lowest point in a local area of the graph.

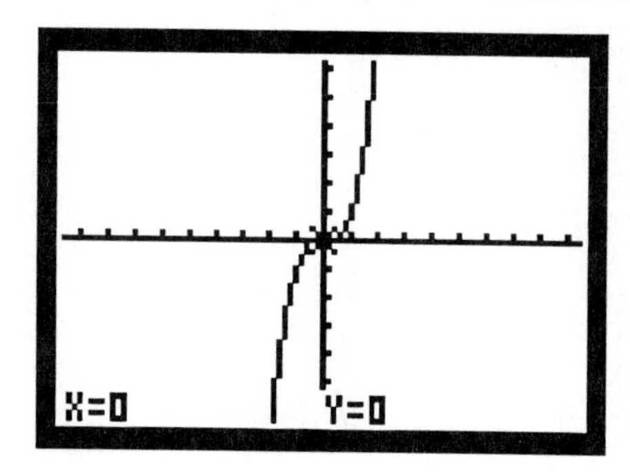

FIGURE 3.19

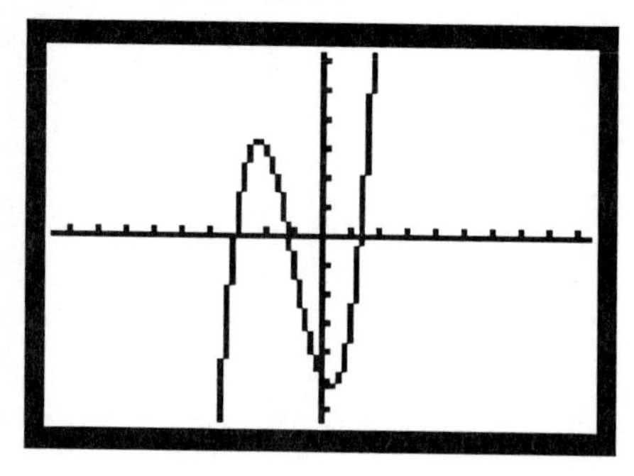

FIGURE 3.20

Figure 3.19 shows the parent function $y = x^3$ drawn in the [-9.6, 9.4] by [-6.4, 6.2] viewing window. This function has only one real root since it crosses the x-axis only once. Use the TRACE cursor to confirm that $f(x) = 0$ when $x = 0, f(x) \to \infty$, as $x \to \infty$ and $f(x) \to -\infty$ as $x \to -\infty$. As the TRACE cursor moves across the graph of the function from left to right (x values increasing), the function values never decrease. This means that this function has no local maximum or minimum. In general, we say this function has no turning point. Is the same true for the function $y = -x^3$?

Figure 3.20 shows the graph of $y = x^3 + 3x^2 - 2x - 5$. This cubic function has three real roots, corresponding to the three points where the graph crosses the x-axis. Using zoom in with error at most .01, the roots of the cubic are $x = -3.128$, $x = -1.202$, and $x = 1.330$. Confirm these values by evaluating the function at each of the points (see Figs. 3.21 and 3.22). The y-intercept of this function is the point (0, -5). Why?

The function has two turning points; a local maximum and a local minimum. Using zoom in with error at most .01, the local maximum point is (-2.291, 3.303)

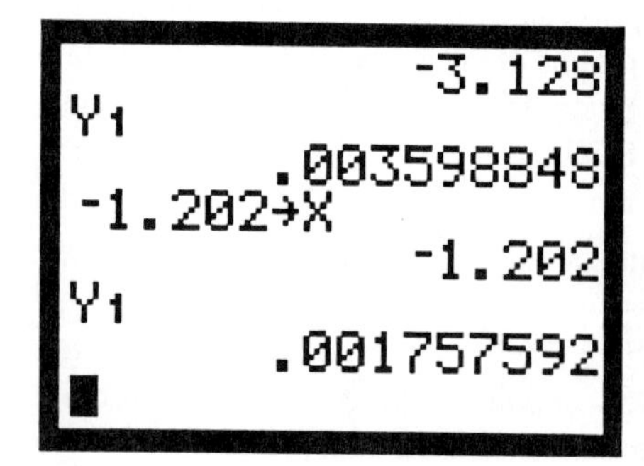

FIGURE 3.21

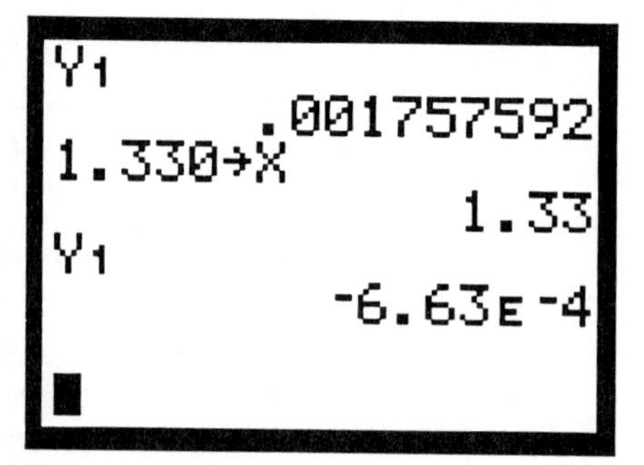

FIGURE 3.22

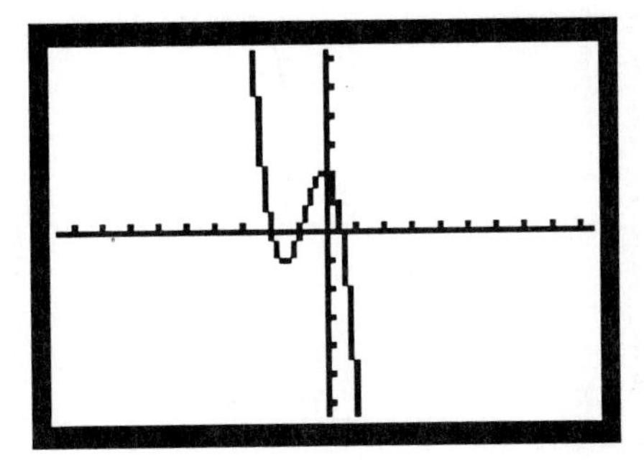

FIGURE 3.23

and the local minimum point is (.291, -5.303). The function is increasing on the intervals

$$-\infty < x < -2.291 \quad \text{and} \quad .291 < x < \infty.$$

Another notation for this solution is union of sets:

$$(-\infty, -2.291) \cup (0.291, \infty).$$

The function is decreasing on the interval

$$(-2.291, .291).$$

Figure 3.23 shows the graph of the function $y = -2x^3 - 5x^2 - x + 2$. Compare the graph of this function to the graph of $y = -x^3$ (see Fig. 3.24). For the function $f(x) = -x^3$, $f(x) \to \infty$ as $x \to -\infty$ and $f(x) \to -\infty$ as $x \to \infty$. This is the opposite of the parent function $f(x) = x^3$. The sign of the leading coefficient gives an indication of the end behavior of the cubic function, just like for quadratic functions.

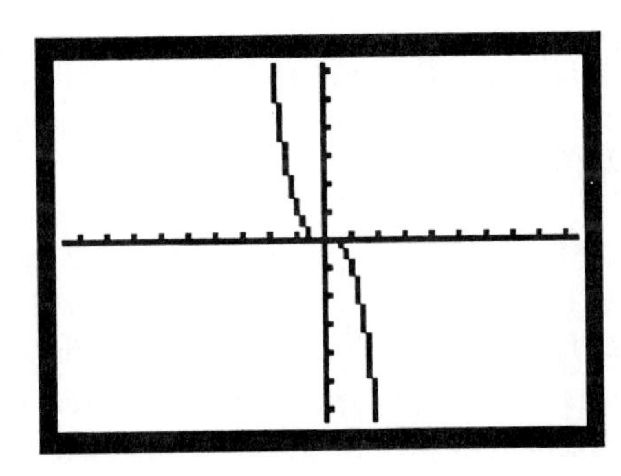

FIGURE 3.24

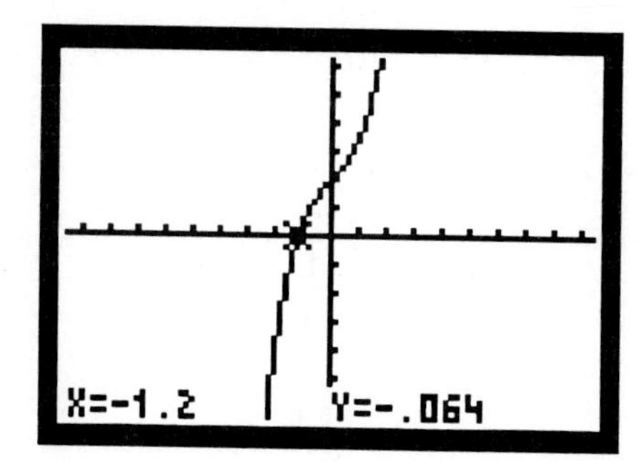

FIGURE 3.25

The function $y = -2x^3 - 5x^2 - x + 2$ has two turning points at (-1.560, -1.015) and (-.107, 2.052). The function is increasing on the interval (-1.560, -.107) and is decreasing on the intervals $(-\infty, -1.560)$ and $(-.107, \infty)$. The y-intercept is the point (0, 2).

This function also has three real roots at $x = -2$, $x = -1$, and $x = .5$. Use the TRACE cursor and zoom in to confirm these values. We could write the function in factored form by solving each of the roots for zero:

$$\begin{array}{lll} x = -2 & x = -1 & x = 0.5 \\ x + 2 = 0 & x + 1 = 0 & 0 = 0.5 - x \\ & & 0 = 1 - 2x \end{array}$$

Then we can say that:

$$-2x^3 - 5x^2 - x + 2 = (x + 2)(x + 1)(1 - 2x).$$

To confirm that these two expressions are equivalent, expand the right side to get the left side:

$$(x + 2)(x + 1)(1 - 2x) = (x^2 + 3x + 2)(1 - 2x) = -2x^3 - 5x^2 - x + 2.$$

Figure 3.25 shows the graph of $y = .5x^3 + x + 2$. The function has no turning points since it is increasing across the entire domain (x values). The function has only one real root at $x = -1.179$. The y-intercept is the point (0, 2).

GENERAL STATEMENT: In these examples we have seen cubic functions of the form $f(x) = ax^3 + bx^2 + cx + d$ $(a \neq 0)$ with one or three real roots and zero or two turning points. From our previous work, we know that quadratic functions of the form $f(x) = ax^2 + bx + c$ $(a \neq 0)$ have zero or two real roots and one turning point, and linear functions of the form $f(x) = ax + b$ $(a \neq 0)$ have one real root and no turning point. Table 3.1 summarizes these results.

TABLE 3.1
NUMBER OF REAL ROOTS AND TURNING POINTS FOR
nth DEGREE POLYNOMIALS

Degree	*Form*	*Real Roots*	*Turning Points*
1	$f(x) = ax + b$	1	0
2	$f(x) = ax^2 + bx + c$	0 or 2	1
3	$f(x) = ax^3 + bx^2 + cx + d$	1 or 3	2
4			
5			
6			
7			
8			
n			

Make a conjecture about the number of real roots and turning points for a fourth degree polynomial function, and then confirm your guess by drawing graphs of many different examples. Continue this exploration for other higher degree polynomial functions and complete the table.

The intervals where a function is increasing or decreasing are directly related to the number of turning points and the sign of the leading coefficient. Visual inspection of the graph of the function together with the values of the turning points will help define the specific intervals. ◊

APPLICATION EXPLORATION: Susan Perez is a packaging engineer working for the Gourmet Nut Company. Her assignment for today is to design packages for their new line of fancy giant cashews that will be marketed in three sizes: small, medium, and large. All of the packages will be constructed from the same 40 by 60 cm rectangular sheets of cardboard by cutting squares from the corners and folding the sides up to form a box with no top. As her first experiment, Susan estimates that the small size will hold 5000 cm^3, the medium size will hold 7500 cm^3, and the large size will hold 10,000 cm^3. Find the dimensions of the three different packages, determine if this plan is feasible, and make any necessary adjustments.

Figure 3.26 shows a drawing of the rectangular cardboard and the planned cuts to be made.

The volume of the finished boxes will be the product of the length, width, and height, or $V = lwh$. Since the corner cutouts are squares, the height of the finished box will be x, the length of the cut. Both the length and the width of the finished boxes will be reduced by $2x$; we can say that *length* $= 60 - 2x$ and *width* $= 40 - 2x$. Volume as a function of x, the length of the cut, is given by

$$V(x) = (60 - 2x)(40 - 2x)x$$

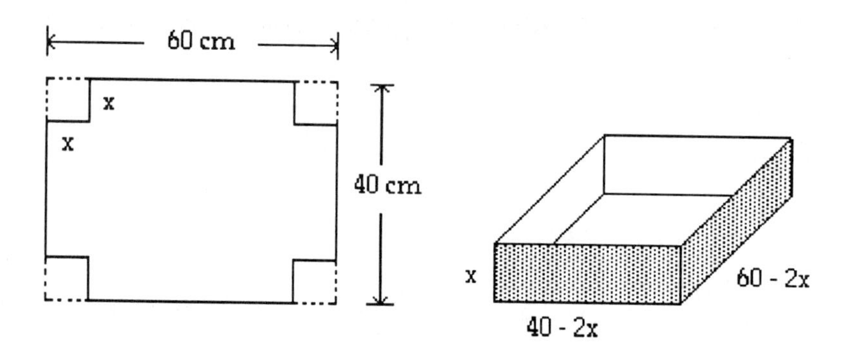

FIGURE 3.26

or, by expanding,

$$V(x) = 4x^3 - 200x^2 + 2400x.$$

The questions to answer are what lengths of x, the cut in the corners, will produce boxes of volume 5000, 7500, and 10,000 cm^3. These questions stated as equations are

$$\begin{aligned}(60-2x)(40-2x)x &= 5000, \\ (60-2x)(40-2x)x &= 7500, \\ (60-2x)(40-2x)x &= 10{,}000,\end{aligned} \quad \text{or} \quad \begin{aligned}4x^3 - 200x^2 + 2400x &= 5000, \\ 4x^3 - 200x^2 + 2400x &= 7500, \\ 4x^3 - 200x^2 + 2400x &= 10{,}000.\end{aligned}$$

First, graph the function $y = (60 - 2x)(40 - 2x)x$. To set the RANGE, examine values that make sense in the problem situation. Since this is a cubic function, it can have up to three real roots. We can find the roots by setting each factor in $V(x) = (60 - 2x)(40 - 2x)x$ equal to zero and solving:

$$\begin{aligned}60 - 2x &= 0 \\ x &= 60 \\ \mathbf{x} &\mathbf{= 30}\end{aligned} \qquad \begin{aligned}40 - 2x &= 0 \\ 2x &= 40 \\ \mathbf{x} &\mathbf{= 20}\end{aligned} \qquad \mathbf{x = 0}$$

This means that the graph of $V(x) = (60 - 2x)(40 - 2x)x$ will pass through the x-axis at the points (0, 0), (20, 0) and (30, 0). Set the domain to:

Xmin = -5, Xmax = 40, Xscl = 5.

To get some idea of the vertical RANGE, choose several x values that make sense in the problem, and evaluate the function at these points. Figure 3.27 shows the values $x = 9$ and $x = 10$ evaluated on the **home screen** ($f(9) = 8316$ and $f(10) = 8000$). We also know that Susan plans to put 10,000 cm^3 of cashews in the large size box. Based on these facts, set the vertical RANGE to:

Ymin = -3000, Ymax = 11,000, Yscl = 1000.

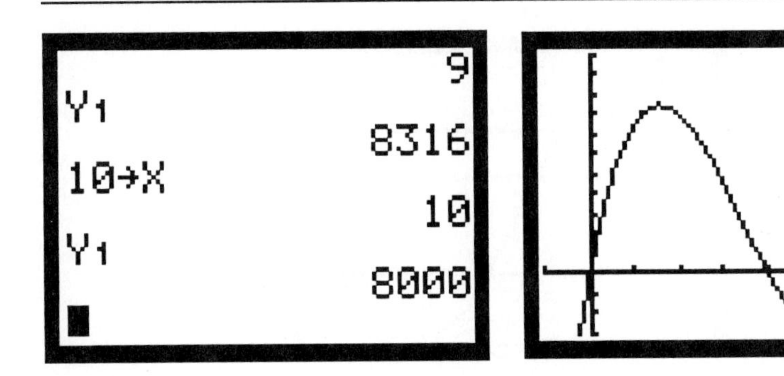

FIGURE 3.27

FIGURE 3.28

Figure 3.28 shows the complete graph of the mathematical function in the window [-5, 35] by [-3000, 11,000]. This graph shows more information than a graph of the problem situation. Since x represents the length of the cut, the only values that make sense in the problem are $0 \leq x \leq 20$. (Why?) In the problem, the function values represent volume, so the only values that make sense are $0 \leq y \leq 10{,}000$. Figure 3.29 shows the graph of the problem situation in the viewing window [0, 20] by [0, 10,000].

It is very important to understand that the graph of the problem situation is a subset of the graph of the complete function. Most of the time it is important to look at the complete graph as well as the graph of the problem situation to fully understand the problem.

To answer Susan's first question, enter a second function $\mathbf{Y_2 = 5000}$, and draw both graphs at the same time. The points of intersection between the two graphs are the answers to this question (see Fig. 3.30). Use zoom in to find the value for x, the length of the cut to produce a 5000 cm^3 box:

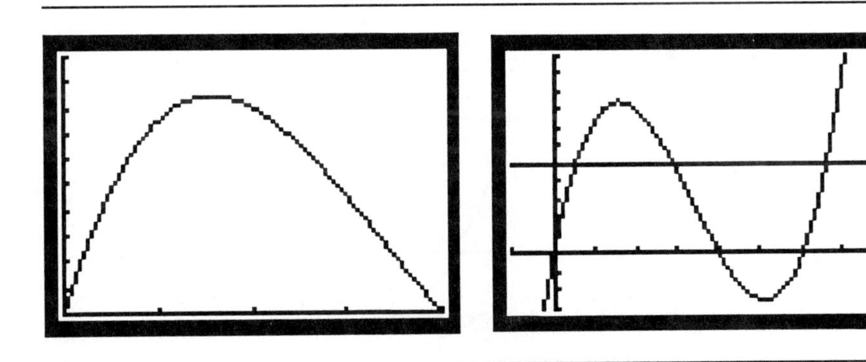

FIGURE 3.29

FIGURE 3.30

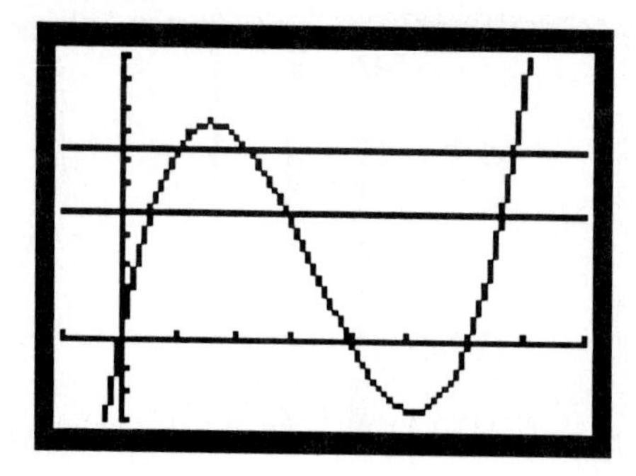

FIGURE 3.31

$x = 2.629, \qquad x = 14.437, \qquad \text{or} \qquad x = 32.934$

Even though there are three intersections between the line and the cubic function, only two of these values make sense in the problem, $x = 2.629$ cm and $x = 14.437$ cm. A cut of 32.934 cm in from all the sides of the cardboard sheet will be more than the 40 cm width.

Add a third equation, **Y_3 = 7500,** to answer the next question about the medium size box (see Fig. 3.31). Again, only two intersections make sense in the problem situation, $x = 5$ cm and $x = 11.044$ cm. The third intersection at $x = 33.956$ is not a possible cut size.

Add a fourth equation, **Y_4 = 10,000,** to answer the final question about the large size box (see Fig. 3.32). This line does not intersect the cubic graph in the domain of the problem situation. The only intersection is at $x = 34.837$, a value that does not make sense in the problem. This means that Susan cannot make a rectangular box with a volume of 10,000 cm^3 from a sheet of cardboard 40 by 60 cm!

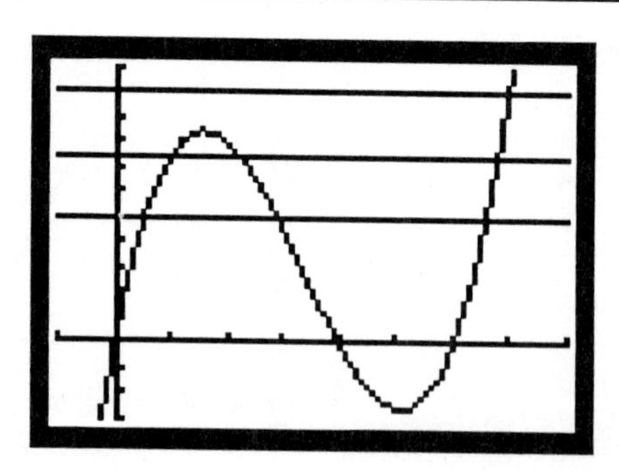

FIGURE 3.32

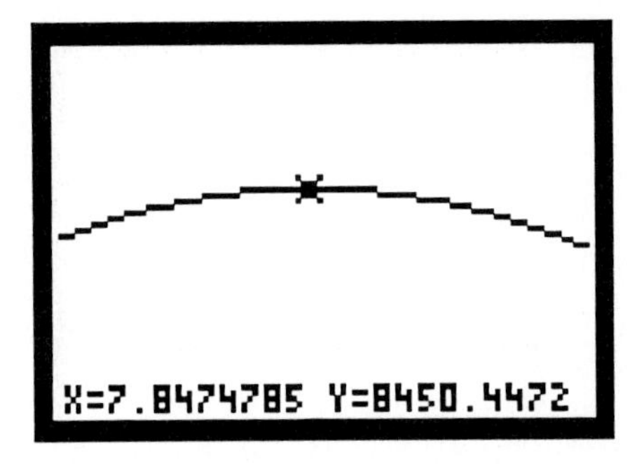

FIGURE 3.33

The volume of the largest box that can be made from the 40 by 60 cm sheet of cardboard can be found by zooming in on the local maximum point in the domain of the problem situation. Figure 3.33 shows this point after zooming in several times using the **Box** command. The largest box that can be cut from the 40 by 60 cm cardboard is 8450.447 cm^3, when squares 7.847 cm on a side are cut from the corners.

At this time, Susan has some decisions to make. She could call this largest box the large size and not change the volume of the small and medium boxes. Or, she could use the maximum size as a guide and readjust the other sizes proportionally. Originally, the small size box was 50% of the volume of the large size box and the medium box was 75% of the volume of the large size box. Using these ratios, the large box would be 8450.447 cm^3, the medium box would be 6337.835 cm^3, and the small box would be 4225.224 cm^3. Using these new values, recalculate the size of the cut for the small and medium size boxes.

Alternate Solution Strategy

Throughout this example we have been solving a system of equations for an intersection. For example, we solved the equations

$$(60 - 2x)(40 - 2x)x = 5000$$

by finding the intersections of the two functions $y = (60 - 2x)(40 - 2x)x$ and $y = 5000$.

Another way to arrive at the same solutions is to solve the original equations for zero and then find the x-intercepts of the resulting equation:

$$(60 - 2x)(40 - 2x)x - 5000 = 0.$$

Instead of reading the answer at the intersection of two functions, the same values occur on the x-axis. From a translational perspective, the original function is shifted 5000 units downward. Figure 3.34 shows the graphs of

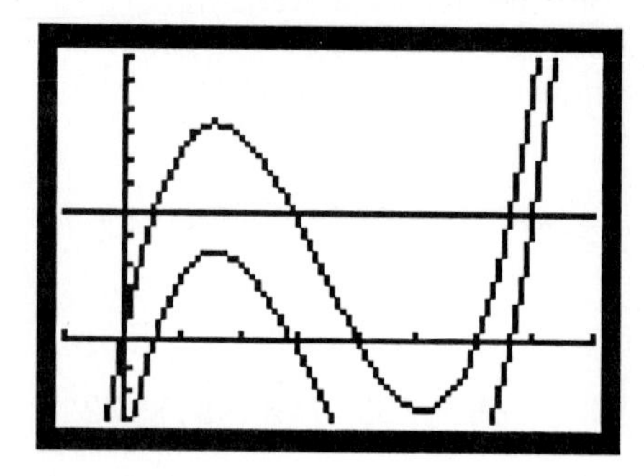

FIGURE 3.34

$$Y_1 = (60 - 2X)(40 - 2X)X,$$
$$Y_2 = 5000,$$
$$Y_3 = (60 - 2X)(40 - 2X)X - 5000,$$
$$(\text{or } Y_3 = Y_1 - Y_2).$$

in the [-5, 40] by [-3000, 11,000] viewing window. Check numerically, using zoom in, that the roots of the third equation are identical to the x values of the intersections of the first two equations.

What happens to the local maximum value for the function

$$Y_3 = (60 - 2X)(40 - 2X)X - 5000?$$

Since the translation is vertical only, the x value of the maximum point is the same for both graphs. However, the y value, representing the maximum volume of the box, is 5000 cm^3 too low. Care should be taken when using this solution strategy to find maximum and/or minimum points. ◊

Problems

1. Graph $y = x$, $y = x^2$, and $y = x^3$ together. Algebraically explain any similarities and differences.
   - a. When x is close to zero, how are they alike?
   - b. When x is positive and large, how do they differ?
   - c. Why does $y = x^3$ increase faster? What happens in the interval $-1 \le x \le 1$? Graph $y = x^3$ in a [-1, 1] by [-1, 1] window.
   - d. When x is negative, how does $y = x^2$ differ from the other functions?
   - e. Why would $y = x^2$ stay positive when the others are negative?

2. From your observations in the previous problem, can you predict how the graphs of $y = x^4$ and $y = x^5$ will behave? Check your predictions by graphing them with $y = x^2$ and $y = x^3$. What happens for $-1 < x < 1$? Can you make a generalization for higher odd and even degree polynomials? How can you explain the behavior algebraically?

3. From your work with different transformations of $y = x^2$, can you predict the graphs of the following transformations of $y = x^3$? Check by graphing them.

   a. $y = x^3 + 3$
   b. $y = x^3 - 4$
   c. $y = (x - 1)^3$
   d. $y = (x + 2)^3$
   e. $y = 3x^3$
   f. $y = -x^3$
   g. $y = (2 - x)^3$
   h. $y = 3(x - 1)^3 + 2$

4. We know what happens if we add a constant to any function, but what if we add the functions $y = x^3$, $y = x^2$, and $y = x$ together?

   a. Graph $y = x^3 + x^2$ in a standard window. Why does this function behave like $y = x^3$ for large values of x? Algebraically, does the term x^3 or x^2 "dominate" for large positive x? Which function dominates for large negative values of x?

   Graph $y = x^3 + x^2$ in the smaller window, for example [-3, 3] by [-2, 2]. Why does the curve cross the x-axis at $x = -1$? Why is there a change in direction between $x = -1$ and $x = 0$? Can you explain this change in behavior numerically for small values of x? For small values of x (ie. $-1 < x < 1$) does the term x^3 or x^2 "dominate"? Which is larger for the interval $-1 < x < 0$? For the interval $0 < x < 1$? Graph $y = x^3 + bx^2$ in a **Standard** window and experiment by increasing the value b. Explain the effect of b on the shape and position of the graph.

   b. Find the x-intercepts for $y = x^3 + x^2$ by factoring. Knowing the general shape of a positive cubic, what should the graph of $y = x^3 - x^2$ look like? Make a prediction of the shape of the graph of $y = x^3 - bx^2$ for different values of b. Does the fact that $x^2 > x^3$ on the interval $-1 < x < 1$ influence the turning point of the graph?

   c. Graph $y = x^3 + x$. Why didn't we get a change in direction (turning point) for this graph? Try graphing $y = x^3 - x$ and explain the difference algebraically. Compare the terms x^3 and $-x$ for $x < 0$. Which function is larger for $-1 < x < 0$? Why is the graph of $y = x^3 + x$ shaped like the graph of $y = x^3$ for large x values?

   d. Graph $y = x^3 + x^2 + x$, and explain the curve numerically. What happens to the shape and position of the graph if we increase the value of the coefficient of x^2 and graph $y = x^3 + 2x^2 + x$? What if we increase the coefficient of x in the same way?

   e. What happens to the graph if the coefficient of the x^2 term is negative? What if the coefficient of the x term is negative? What if they are both negative? Explain.

5. What happens to the shape and position of the graph of the simple cubic $y = x^3$ if it is added to a known quadratic? Compare the graph of the cubic $y = x^3 + ax^2 + bx + c$ to the quadratic $y = ax^2 + bx + c$ for the following functions. Will they always match up this way? Explain why some match up better than others.

   a. $y = x^3 + 3x^2 + 2x + 1$ and $y = 3x^2 + 2x + 1$
   b. $y = x^3 + 3x^2 + 5x - 2$ and $y = 3x^2 + 5x - 2$
   c. $y = x^3 + 6x^2 - 5x + 2$ and $y = 6x^2 - 5x + 2$
   d. $y = x^3 + x^2 + x + 1$ and $y = x^2 + x + 1$

6. Investigate the standard cubic $y = ax^3 + bx^2 + cx + d$ by graphing it for different values of a, b, c, and d. Then answer the following:
   a. How does changing a affect the graph of the function? What if $a < 0$?
   b. How does changing d affect the graph of the function? What if $d < 0$?
   c. How does changing b affect the graph of the function? Try larger values of b. What if $b < 0$?
   d. What is the effect of c? Graph $y = ax^3 + bx^2$ with $y = ax^3 + bx^2 + cx$ to help see the effect of c. What if c is larger? What if $c < 0$?
7. We have already shown that the product of two lines is a parabola. What is the product of a line and a parabola? In a friendly window, graph the line $\mathbf{Y_1 = X}$, the parabola $\mathbf{Y_2 = X^2}$, and their product $\mathbf{Y_3 = X * X^2}$ (or $\mathbf{Y_3 = Y_1 * Y_2}$). Now TRACE between the three functions to convince yourself that the product of the first two is the third, for each value of x. Explain algebraically the change of sign for the product of the two functions. Explain why this cubic function is negative for negative values of x.
8. Repeat the same experiment from problem 7 with $\mathbf{Y_1 = X - 3}$, $\mathbf{Y_2 = X^2 - 4X - 6}$, and $\mathbf{Y_3 = (X - 3)(X^2 - 4X - 6)}$ in the window [-9, 10] by [-15, 25]. Explain how the y value of the cubic changes sign in relation to the signs of the linear and quadratic functions. Explain what happens when one is positive and the other is negative; when both are positive; when both are negative.
9. Graph and explore the product $\mathbf{Y_3 = (X - 3)(X^2 + 4X - 2)}$. To cross the x-axis twice, there must be a change in direction (a turning point). Explain numerically how the change of direction for the cubic function in the first quadrant occurs.
10. Explore what happens when we multiply two parabolas. To investigate this, graph $\mathbf{Y_1 = 2X^2 - X - 3}$, $\mathbf{Y_2 = -X^2 + 2X + 1}$, and $\mathbf{Y_3 = Y_1Y_2}$ in a friendly window. To see this product better, move the TRACE cursor to integer y values.
11. Up to now all of the polynomials that we have graphed could be visualized in the **standard** window. Use what you have learned so far to find a window that contains the complete graph of each of the following polynomials:
   a. $y = x^3 + 5x^2 + 6$
   b. $y = x^3 + 11x^2 - 11x - 18$
   c. $y = 3x^3 - 3x^2 - 4x - 15$
   d. $y = x^4 - x^3 - x^2 - x - 18$
   e. $y = x^4 + 3x^3 - 3x^2 - 4x - 15$
   f. $y = x^4 + 12x^3 + 2x^2 - x - 1$
12. Because graphing calculators are discrete machines, what we see may not be an accurate picture of the function. This is sometimes caused by "hidden" behavior. Use ZOOM to check for any "hidden" behavior for the following polynomials:
   a. $y = 3x^3 - 3x^2 + 1$ (how many roots?)
   b. $y = x^3 + 3x^2 - x$ (how many roots?)
   c. $y = x^5 - 8x^4 + 25x^3 - 38x^2 + 28x - 8$ (how many roots?)
   d. $y = x^3 + 2x^2 + x - 1$ (any turning points?)
   e. $y = x^3 - 2x^2 + x - 3$ (any turning points?)
   f. $y = 20x^3 + 2x^2 - 1$ (any turning points?)
13. The ABCD Company (All the Best Compact Discs) manufactures CDs. The monthly supply is determined by the function $S = 1 + .1x^3$, and the monthly demand is determined by $D = 750 - x^2$, where x is the price of each CD. When are the supply and demand equal? Find both the price and the number of CDs produced.

14. Cut equal squares out of the corners of an 8.5 by 11 in. piece of paper, and fold up the sides to form a box without a top.

a. Using x as the side of the square cut from each corner, write the volume as a function of x.

b. Graph the function from part (a). What values of x make sense in this problem? Explain.

c. What is the volume if the length of the side of the square cut out is 1 in.?

d. What size square will give you a volume of 40 in.^{3}?

e. What size square gives the maximum volume of the box? What is the maximum volume?

15. Graphically investigate the following identities to be sure they are true statements for all values of x, and check algebraically. Is there any pattern for the coefficients?

a. $(x + 1)^2 = x^2 + 2x + 1$

b. $(x + 1)^3 = x^3 + 3x^2 + 3x + 1$

c. $(x + 1)^4 = x^4 + 4x^3 + 6x^2 + 4x + 1$

d. $(x + 1)^5 = x^5 + 5x^4 + 10x^3 + 10x^2 + 5x + 1$

16. Use a friendly integer window and TRACE to find three consecutive integers whose product satisfies each of the following conditions in turn.

a. A four-digit number such that the first two digits and the second two digits are consecutive numbers.

b. A four-digit number such that the first two digits are double the second two digits or vice versa.

c. A six-digit number such that the first three digits are equal to the second three digits.

d. A seven-digit number such that the first three digits are equal to the last three digits.

e. Can you find any more patterns dealing with three or more consecutive numbers and their product?

17. Investigate the following functions by graphing in an integer window. In the following functions, n represents an integer and "divisible" means that the quotient is an integer. (*Hint*: Try $n = 1$, $n = 2$, or $n = 3$.)

a. $y = \dfrac{x^3 - x}{n}$ is divisible for which single-digit integer values of n?

b. $y = \dfrac{x^5 - x}{n}$ is divisible for which positive integer values of n?

18. In the previous chapter we discovered that the turning point (vertex) of a standard quadratic function $y = ax^2 + bx + c$ is the point $\left(\dfrac{-b}{2a}, c - \dfrac{b^2}{4a}\right)$. Graph a quadratic formed using x^2 ($y = a(x^2)^2 + bx^2 + c$) and one using x^3 ($y = a(x^3)^2 + bx^3 + c$) along with the standard quadratic $y = ax^2 + bx + c$ for the same values of a, b, and c. Trace in a friendly window to answer the following questions.

a. Explain what effect a negative value for b has on each graph.

b. What do you notice about the y-coordinate of the turning points?

c. Can you make any conjecture about the x-coordinate of the turning points and the value $\dfrac{-b}{2a}$? Try $a = 1$ and $b = -2$. Try $a = 1$ and $b = -8$. Try $a = 1$ and $b = -16$.

d. If the x-coordinate of the turning point of $y = ax^2 + bx + c$ is $x = \frac{-b}{2a}$, of $y = a(x^2)^2 + bx^2 + c$ is $x = \sqrt{\frac{-b}{2a}}$, and of $y = a(x^3)^2 + bx^3 + c$ is $x = \sqrt[3]{\frac{-b}{2a}}$, then express in words the relationship between the function in the quadratic and this value. How does this explain the number of turning points?

FOR ADVANCED ALGEBRA STUDENTS

19. Complete Table 3.1. For the following functions, predict the number of real zeros and turning points, and then draw the graphs to check your guesses.

a. $y = x^4 + 2x^3 - 5x^2 - 3x + 7$

b. $y = x^5 + x^4 - 9x^3 + 2x^2 + 6x - 5$

c. $y = x^6 + 5x^5 - 4x^4 - 10x^3 + 6x^2 + x - 5$

d. $y = x^7 + 4x^6 - 14x^5 - 56x^4 + 49x^3 + 196x^2 - 36x - 144$

20. We have seen quadratics that have two real roots or no real roots. Using transformations, make up equations with the following properties. If it is impossible, explain why.

a. A cubic with three real roots. Two. One. None.

b. A fourth degree with four real roots. Three. Two. One. None.

c. A third degree polynomial with two turning points. One. None.

d. A fourth degree polynomial with one turning point. Two. Three. Four. None.

21. To solve the quadratic $2x^2 - x - 6 = 0$, you factor to get $(2x + 3)(x - 2) = 0$, and you get rational roots of $x = \frac{-3}{2}$ and $x = 2$. Rational roots are always dependent on the factors of the lead coefficient (highest degree term) divided into the factors of the constant term (lowest degree term). Enter the following polynomials onto the Y= menu, and use the same rules as you do for quadratics to list the possible rational roots. Graph to estimate which ones may be rational roots. To test them, use the STO key to enter them into the **X** memory register, and then evaluate $\mathbf{Y_1}$. What value returned for $\mathbf{Y_1}$ tells you that you have found a root of the equation?

a. $x^3 + 6x^2 + 11x + 6 = 0$

b. $6x^3 - 25x^2 + 31x - 10$

c. $6x^4 - 7x^3 - 23x^2 + 14x + 24 = 0$

d. $x^5 + 3x^4 - 5x^3 - 15x^2 + 4x + 12 = 0$

22. For any polynomial function $ax^n + bx^{n-1} + \cdots + k = 0$, the sum of the roots is $\frac{-b}{a}$, and the product of the roots is $\frac{k}{a}$ when n is even and $\frac{-k}{a}$ when n is odd. Check this for the equations in problem 21.

23. In each of the following, write a third degree polynomial that has the given rational roots, and check by graphing it.

a. $x = 1$, $x = 2$, and $x = -3$

b. $x = -1$, $x = 2/3$, and $x = -5/2$

24. Graph the following to investigate what happens when a polynomial has multiple roots. What is the difference between what happens to the graph for a multiple root repeated an odd number of times and one repeated an even number of times? Check this for higher degree polynomials.

a. $y = (x-1)(x-1)(x+2)$
b. $y = (x-1)(x-1)(x-1)(x+1)$
c. $y = x^3 - x^2 - 5x - 3$
d. $y = x^4 + 5x^3 + 6x^2 - 4x - 8$

25. A double root occurs when a curve is tangent to the x-axis instead of crossing through the x-axis. Graphically investigate the following:

a. For what two values of k does $2x^3 - 9x^2 + 12x - k = 0$ have a double root?

b. For what values of k does $4x^3 + kx - 27 = 0$ have a double root?

26. Investigate graphically whether $x^n - 1$ is divisible by $x - 1$ for any natural number n. For $x - 1$ to be a factor, $x = 1$ must be a root.

27. Find the points of intersection for the following systems of equations:

a. $y = x^3 - x - 7$ and $y = -x + 2$
b. $y = x^3 + x^2 - x + 1$ and $y = x^2 - x - 1$
c. $y = x^3 - 3x^2$ and $y = x^3 + 3x^2 - 5x + 6$
d. $y = x^2 - x$ and $y = x - x^2$

28. Solve the following inequalities:

a. $x^4 < x^2$
b. $x^3 + 1 > x^2 + x$
c. $x^3 + x^2 + x + 1 \geq x^3 - x^2 - x - 1$
d. $x^5 - x^3 - 1 \leq x^4 + x^2 + x$

29. For any cubic $y = ax^3 + bx^2 + cx + d$, the x-coordinate of the inflection point (the point where a curve changes concavity) is $x = \frac{-b}{3a}$. Graph the following polynomials and TRACE to see if this seems to be correct.

a. $y = x^3 + x^2 - x - 1$
b. $y = 3x^3 - 2x^2 + 6x - 7$

30. We have used a friendly integer window many times to examine number theory. Use the TRACE cursor to find the following and check both algebraically and numerically.

a. Three consecutive integers such that their product is 800 times their sum.

b. Three consecutive even integers such that their product is 132 times their sum.

c. Three consecutive odd integers such that their product plus 9 more than their sum is a perfect square.

31. We graphed the absolute values of linear and quadratic functions in other chapters. Graph the absolute value of several of the polynomials already graphed, and explain the sharp changes in direction on the x-axis. Where do they always occur?

3.2 Exploring Rational Functions

In the last section we explored polynomial functions and their graphs. In this section we will examine more polynomial functions in a special relationship, as a **quotient of two polynomials.** For example, given the two polynomials

$$h(x) = x^3 + 4x^2 - 3x + 1 \quad \text{and} \quad g(x) = 2x^2 - x - 3,$$

we can create a new function from the **ratio** of these two polynomials:

$$f(x) = \frac{h(x)}{g(x)} = \frac{x^3 + 4x^2 - 3x + 1}{2x^2 - x - 3}.$$

A **rational function** is a function that can be expressed as the quotient of two polynomials in the form $f(x) = \frac{h(x)}{g(x)}$. In this form the rational function is like a fraction with a polynomial in the numerator and the denominator.

Any time we are working with a fraction (a ratio), we must pay attention to the denominator. If the denominator of a fraction is any number other than zero, then the fraction has a value, or position on the number line. For example, the following fractions all have values, even the one where the numerator is zero:

$$\frac{3}{4} \qquad \frac{23}{15} \qquad \frac{0}{4} \qquad \frac{-5}{234}$$

However, if the denominator of a fraction is zero, like $\frac{7}{0}$, we say that the fraction is undefined. A fraction indicates division of the numerator by the denominator. Division by zero is undefined.

We discovered in Section 3.1 that some polynomial functions can have one or more zeros for certain values of x, and some polynomials have no real roots and therefore never have a value of zero for any value of x. When we create a rational expression (a fraction) out of two polynomials, we must pay attention to if and when the denominator takes on the value of zero. These points occur for certain values of x.

EXPLORATION 1: Explore the rational function

$$f(x) = \frac{n(x)}{d(x)} = \frac{x^3 + 2x^2 - x + 6}{2x^2 - x - 3}.$$

Enter the numerator as $\mathbf{Y_1}$ and the denominator as $\mathbf{Y_2}$ (see Fig. 3.35). Figure 3.36 shows a graph of only the denominator, d(x) = $2x^2 - x - 3$, in the viewing window [-4.8, 4.7] by [-3.2, 3.1]. Use the TRACE cursor to find the zeros at x = -1 and x = 1.5.

If we evaluate the numerator and denominator at the values x = -1 and x = 1.5,

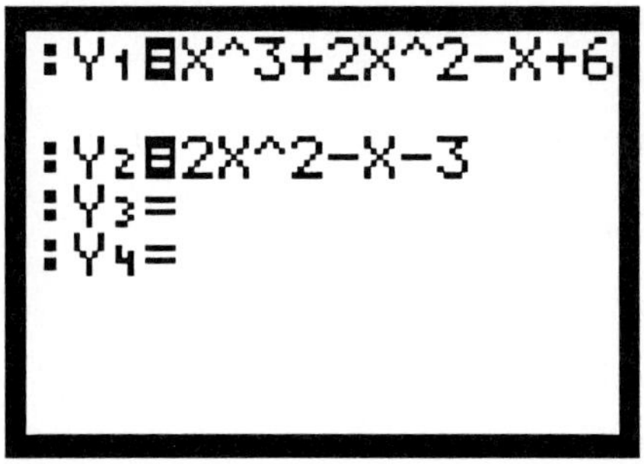

```
:Y1■X^3+2X^2-X+6
:Y2■2X^2-X-3
:Y3=
:Y4=
```

FIGURE 3.35

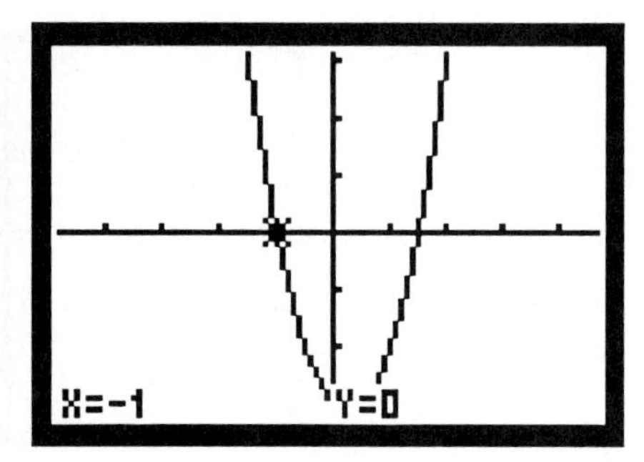

FIGURE 3.36

we get the fractions $\frac{8}{0}$ and $\frac{12.375}{0}$ (see Figs. 3.37 and 3.38). Both of these fractions are undefined for these two values of x.

Based on the graph of $d(x) = 2x^2 - x - 3$, any other values of x will give a non-zero value for the denominator. Why?

What happens if the numerator becomes zero? Figure 3.39 shows a graph of the numerator $n(x) = x^3 + 2x^2 - x + 6$ in the viewing window [-9.6, 9.4] by [-15, 15]. This cubic function has only one real zero, at $x = -3$. If we evaluate both the numerator and denominator at $x = -3$, we get the fraction $\frac{0}{18} = 0$. Figure 3.40 shows the graph of the original rational function $f(x) = \dfrac{x^3 + 2x^2 - x + 6}{2x^2 - x - 3}$ in the viewing window [-9.6, 9.4] by [-6.4, 6.2]. This rational function may be defined on the [Y=] menu in two ways:

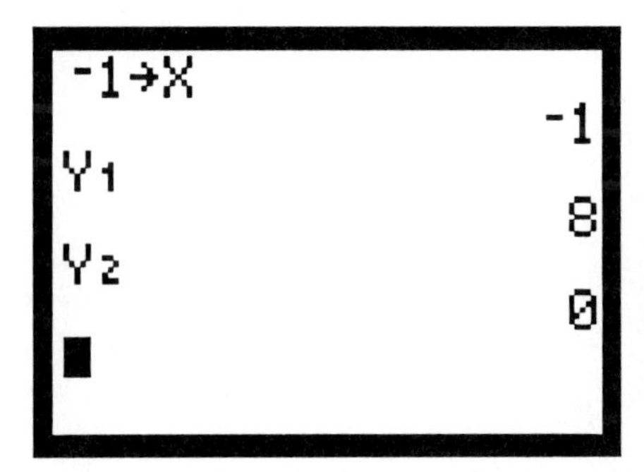

FIGURE 3.37

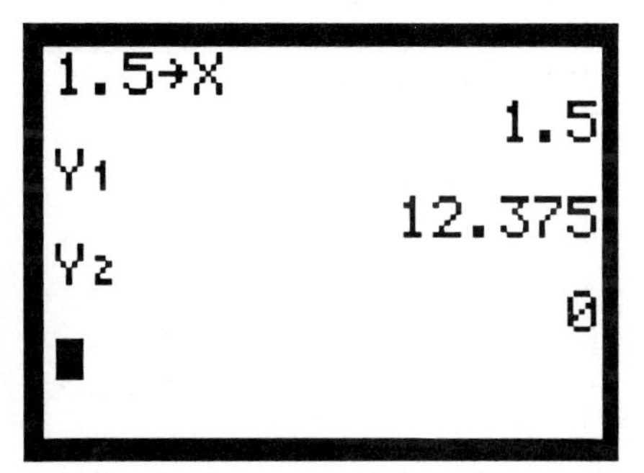

FIGURE 3.38

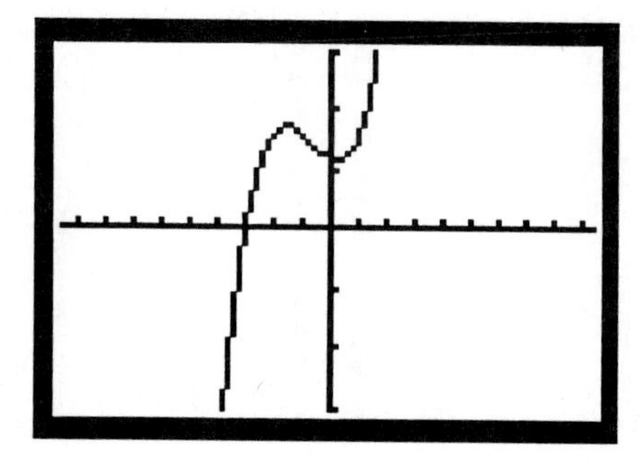

FIGURE 3.39

$$Y_3 = (x^3 + 2x^2 - x + 6)/(2x^2 - x - 3),$$

or, if the numerator and denominator are defined separately,

$$Y_3 = Y_1/Y_2.$$

Now TRACE to the three points $x = -3$, $x = -1$, and $x = 1.5$. What happens to the function values? At $x = -3$, $y = 0$, as we calculated previously. When the numerator of the fraction becomes zero, and the denominator is any nonzero value, the rational function has a real zero, and the graph of the function crosses the x-axis (see Fig. 3.40).

At $x = -1$, there is no value given for the function (see Fig. 3.41). This value is a **point of discontinuity** in the graph. If you tried to trace along the complete graph of this discontinuous function with a pencil, you would have to lift your pencil from the paper to complete the trace. This is not true for the graphs of polynomials; they are continuous functions.

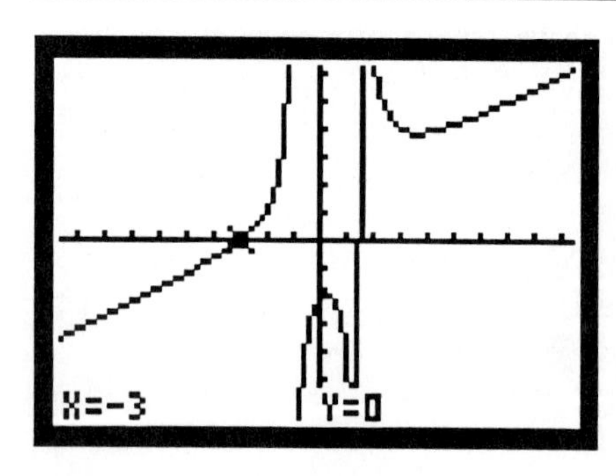

FIGURE 3.40

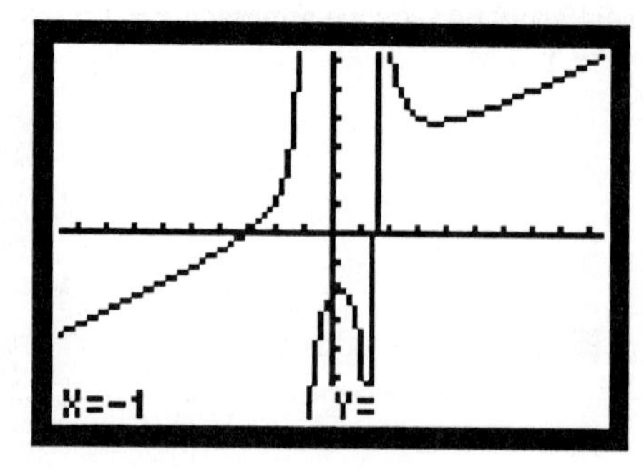

FIGURE 3.41

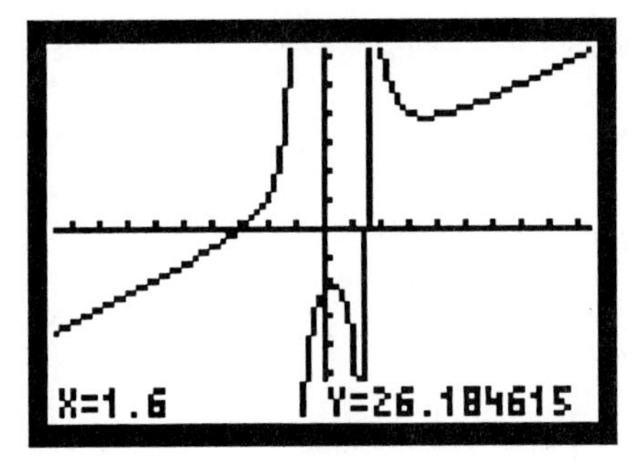

FIGURE 3.42

In the viewing window with the horizontal dimension of [-9.6, 9.4], the exact value $x = 1.5$ is not one of the points evaluated (see Fig. 3.42). Use the TRACE cursor to step across the graph near $x = 1.5$. Notice that the function values change from negative to positive as the TRACE cursor steps from $x = 1.4$ to $x = 1.6$. The vertical line that appears to connect the two branches of this graph is really **not** part of the graph, but only the calculator's attempt to connect points in order. (Your calculator should be in **Connected** mode.) As we know, this function is undefined at $x = 1.5$, so there is a point of discontinuity at this x value.

We did not see this same false vertical line at the point $x = -1$, because in the given viewing window $x = -1$ was a point that the calculator attempted to evaluate. It was undefined for this function, so the calculator did not attempt to connect the points.

Evaluate the rational function at x-values close to the points of discontinuity by placing values in the **X** memory register and evaluating the rational function (stored in the function location $\mathbf{Y_3}$ in this example). Do this for values slightly larger and slightly smaller than the points of discontinuity. Table 3.2 shows this process for the rational function near the value $x = 1.5$.

As the values approach $x = 1.5$ from below, the function values approach $-\infty$ (-247,498.64). As we approach $x = 1.5$ from above, the function values approach $+\infty$ (247,501.36). We can see this on the graph. Below $x = 1.5$, the graph turns sharply downward; above $x = 1.5$, the graph turns sharply upward. Complete a similar table (Table 3.3) for values near $x = -1$ to investigate the behavior of this function near this other point of discontinuity.

If we drew a vertical line through the points $x = -1$ or $x = 1.5$, the graph would approach these lines but never reach them. These lines are called **vertical asymptotes**; they occur at undefined points of a rational function where the denominator of the function becomes zero and the numerator is nonzero. This behavior at a vertical asymptote is called an **infinite discontinuity** because the function value approaches $\pm\infty$ as the x values approach the asymptotes.

TABLE 3.2
FUNCTION VALUES FOR $f(x) = \dfrac{x^3 + 2x^2 - x + 6}{2x^2 - x - 3}$ NEAR $x = 1.5$

x	$f(x) = \dfrac{x^3 + 2x^2 - x + 6}{2x^2 - x - 3}$
1.49	-246.1475703
1.499	-2473.640756
1.4999	-24,748.64008
1.49999	-247,498.64
1.5	undefined
1.50001	247,501.36
1.5001	24,751.36008
1.501	2476.360756
1.51	248.8675498

The line seen on the graph in Fig. 3.42 at approximately $x = 1.5$ is not really an asymptote since it is not exactly vertical. You may think of this line as an asymptote; however, remember that an asymptote is not part of the graph, but only an artificial construct to help you interpret the graph of the function. ◊

Some rational functions are discontinuous at a vertical asymptote, and sometimes another type of discontinuity occurs.

TABLE 3.3
FUNCTION VALUES FOR $f(x) = \dfrac{x^3 + 2x^2 - x + 6}{2x^2 - x - 3}$ NEAR $x = -1$

x	$f(x) = \dfrac{x^3 + 2x^2 - x + 6}{2x^2 - x - 3}$
-1.1	
-1.01	
-1.001	
-1.0001	
-1	undefined
-.9999	
-.999	
-.99	
-.9	

EXPLORATION 2: Compare and contrast the behavior of the graphs of the rational functions

$$f(x) = \frac{x^2 - 1}{x + 1} \quad \text{and} \quad g(x) = \frac{x^2 - 2}{x + 1}.$$

Make and test a conjecture about the zeros of the numerator and denominator of a rational function and the corresponding type of discontinuity.

Both denominators are the same linear polynomial. Set the denominator equal to zero to find the x values where discontinuities might occur:

$$x + 1 = 0$$
$$x = -1$$

Both of the functions will be undefined at $x = -1$. Figure 3.43 shows the graph of $f(x) = \frac{x^2 - 1}{x + 1}$ and Fig. 3.44 shows the graph of $g(x) = \frac{x^2 - 2}{x + 1}$, both in the viewing window [-4.8, 4.7] by [-3.2, 3.1]. The two graphs look very different. The TRACE cursor confirms that both graphs are undefined at $x = -1$ by showing no value for y.

The graph of the function $f(x) = \frac{x^2 - 1}{x + 1}$ seems to have a "hole" at the point (-1, -2). Zoom in repeatedly near the point (-1, -2). The hole in the graph disappears after the first zoom. This is because the exact value $x = -1$ is not one of the values evaluated on the screen of the calculator. If you set **Xmin** and **Xmax** such that $x = -1$ is a point evaluated, then the hole will reappear.

As you zoom in, notice that the function does not turn and head off toward $\pm\infty$ at $x = -1$ as the graph of $g(x) = \frac{x^2 - 2}{x + 1}$ does. The function has a **point discontinuity** at

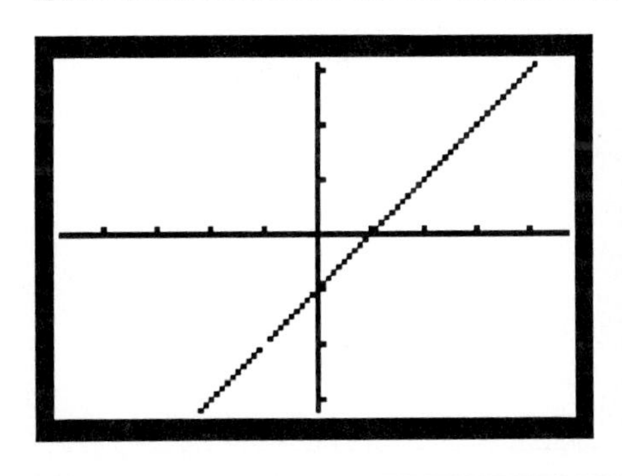

FIGURE 3.43

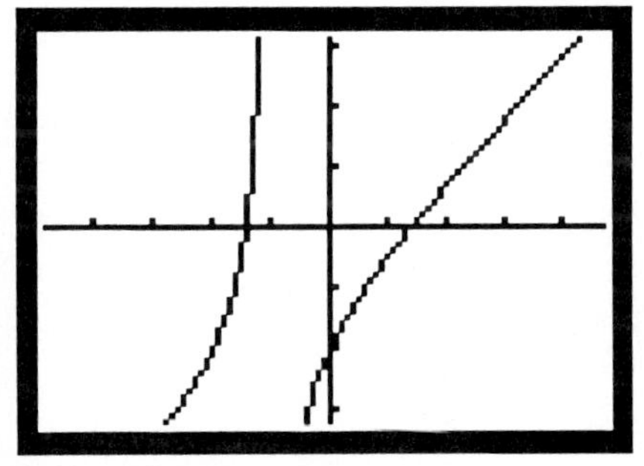

FIGURE 3.44

$x = -1$. This means that there is a single point missing from the graph at $x = -1$. But more than that, if we get close to the value $x = -1$, the point of discontinuity, the function values get closer and closer to a single value. Complete Table 3.4 numerically by using your TI-81 to evaluate points near $x = -1$. What number do the function values seem to be approaching? Is this value the same for x values on both sides of the point of discontinuity? Why? Repeat the same analysis in Table 3.5 for the function $g(x)$ near $x = -1$.

Compare the numerical patterns in Tables 3.4 and 3.5. Are they similar? Does this numerical exploration confirm what the graphs show?

Graphical Exploration

Define the numerator and denominator of the rational functions $f(x) = \dfrac{x^2 - 1}{x + 1}$ and $g(x) = \dfrac{x^2 - 2}{x + 1}$ separately as

$\mathbf{Y_1 = X^2 - 1,}$
$\mathbf{Y_2 = X + 1,}$
$\mathbf{Y_3 = X^2 - 2,}$
$\mathbf{Y_4 = X + 1.}$

Figure 3.45 shows the graphs of $\mathbf{Y_1}$ and $\mathbf{Y_2}$, the numerator and denominator of $f(x)$ in the viewing window [-4.8, 4.7] by [-3.2, 3.1]. What do you notice about the zeros of these two functions?

TABLE 3.4
FUNCTION VALUES FOR $f(x) = \dfrac{x^2 - 1}{x + 1}$ NEAR $x = -1$

x	$f(x) = \dfrac{x^2 - 1}{x + 1}$
-1.1	
-1.01	
-1.001	
-1.0001	
-1	undefined
-.9999	
-.999	
-.99	
-.9	

TABLE 3.5
FUNCTION VALUES FOR $g(x) = \dfrac{x^2 - 2}{x + 1}$ NEAR $x = -1$

x	$g(x) = \dfrac{x^2 - 2}{x + 1}$
-1.1	
-1.01	
-1.001	
-1.0001	
-1	undefined
-.9999	
-.999	
-.99	
-.9	

Figure 3.46 shows the graphs of $\mathbf{Y_3}$ and $\mathbf{Y_4}$, the numerator and denominator of $g(x)$, in the window [-4.8, 4.7] by [-3.2, 3.1]. What do you notice about the location of the zeros of these two functions? Compare these zeros with those shown in Fig. 3.45. The graph of $f(x) = \dfrac{x^2 - 1}{x + 1}$ has a point discontinuity, and the graph of $g(x) = \dfrac{x^2 - 2}{x + 1}$ has an infinite discontinuity.

CONJECTURE: If the numerator and denominator of a rational function have a value of zero at the same x value, then the graph of the function will have a point discontinuity; and if the numerator and denominator have zeros at different x

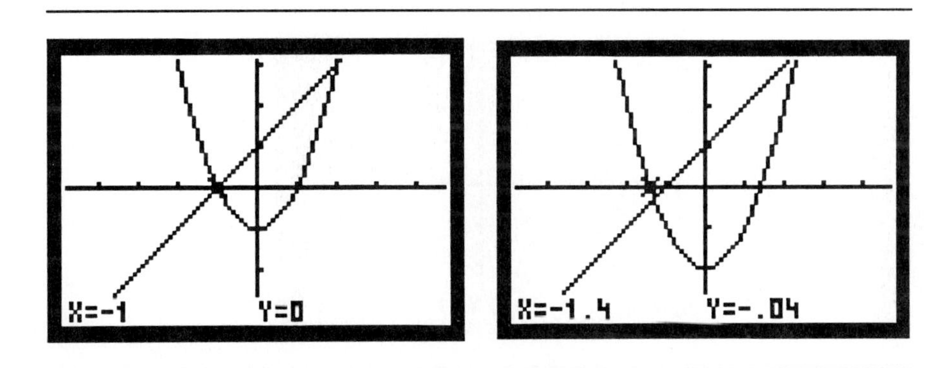

FIGURE 3.45 FIGURE 3.46

values, then the graph of the rational function will have an infinite discontinuity at the x value where the denominator has a zero.

Test this conjecture by making 5 to 10 rational functions from two polynomials that have real zeros. Try to predict the type of discontinuity by examining the zeros of the graphs of the numerators and denominators in pairs. (*Note*: If you can find just one **counterexample,** the conjecture will be false.) ◊

EXPLORATION 3: Investigate the family of rational functions with the parent function $y = \frac{1}{x}$. Use the same procedures for stretching and shifting this parent function as done previously with other families of functions.

Figure 3.47 shows the graphs of $y=\frac{1}{x}$, $y=\frac{2}{x}$, $y=\frac{3}{x}$, and $y=\frac{4}{x}$ in the [-4.8, 4.7] by [-3.2, 3.1] viewing rectangle. These functions are undefined at $x = 0$. Why?

By factoring the numerator of each of these functions, they can be rearranged in the form $y = a\left(\frac{1}{x}\right)$ where a is the vertical stretch factor:

$$y=\frac{1}{x}=1\left(\frac{1}{x}\right),\quad y=\frac{2}{x}=2\left(\frac{1}{x}\right),\quad y=\frac{3}{x}=3\left(\frac{1}{x}\right),\quad y=\frac{4}{x}=4\left(\frac{1}{x}\right)$$

Use the TRACE cursor to confirm that the function values of the parent function are stretched by a factor of 2, 3, or 4 for each of the other functions.

Figure 3.48 shows the graphs of $y=\frac{1}{x}$, $y=.5\left(\frac{1}{x}\right)$, $y=.3\left(\frac{1}{x}\right)$, and $y=.1\left(\frac{1}{x}\right)$. Use the TRACE cursor to confirm that the parent function is compressed by the factors .5, .3, and .1 for each of the other three functions.

These vertical stretching and compressing motions are the same for rational

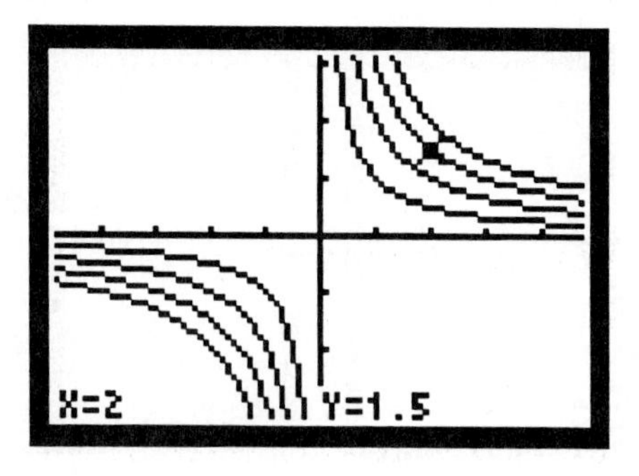

FIGURE 3.47

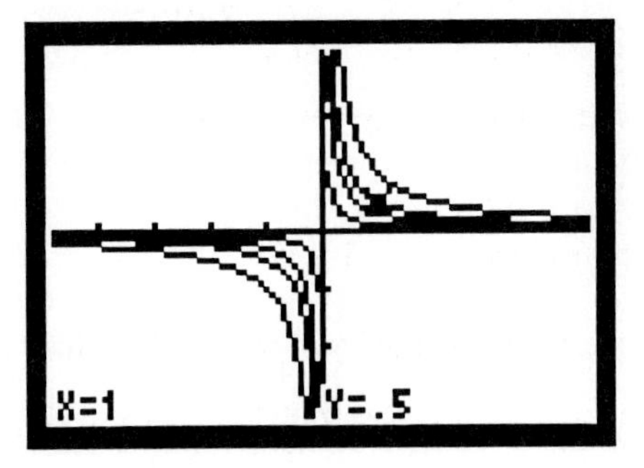

FIGURE 3.48

functions as for other families of functions we have studied. Sometimes a bit of algebraic manipulation is needed to get the function into the transformational format.

Figure 3.49 shows the graphs of $y=\frac{1}{x}$ and $y=\frac{-1}{x}$. As we discovered for other functions, when the stretch factor is negative, the graph of the function is rotated around the x-axis.

Figure 3.50 shows the graph of $y = \frac{1}{x} + 2$, which is the parent function shifted upward 2 units. Use the TRACE cursor to confirm that every point on the parent function is shifted up exactly 2 units. Add the graph of $y = 2$. Notice that the graph of $y=\frac{1}{x} + 2$ seems to approach this horizontal line. When will the graph

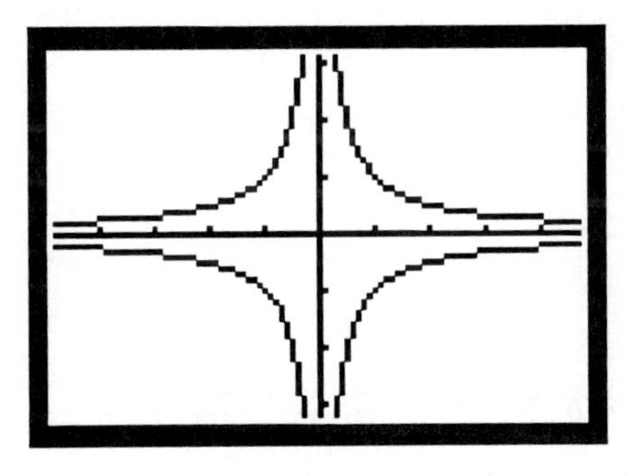

FIGURE 3.49

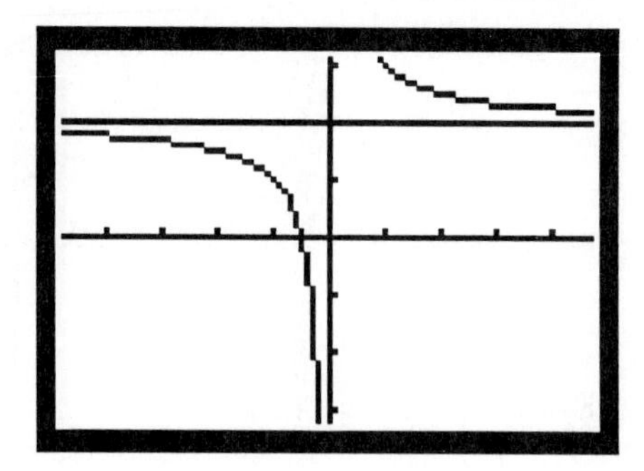

FIGURE 3.50

actually cross this line? Another way to state this question is when will $\frac{1}{x} + 2 = 2$? Move the TRACE cursor along the graph of the function in either direction, and see if it ever equals exactly 2. Or, change **Xmin** and **Xmax** to jump out along the x-axis. Figure 3.51 shows the graph of $y = \frac{1}{x} + 2$ in the viewing window [5000, 6000] by [-3.2, 3.1]. The function is approaching the value $y = 2$, but is not equal to it. Figure 3.52 shows the same function in the window [-10,000, -9000] by [-3.2, 3.1]; the function value still approaches $y = 2$.

We can analyze this question algebraically: When is $\frac{1}{x} + 2 = 2$, or when is $\frac{1}{x} = 0$? The fraction $\frac{1}{x} = 0$ only if the numerator is equal to zero. Since the numerator is not a variable, it will never equal zero. Therefore this fraction will never equal zero,

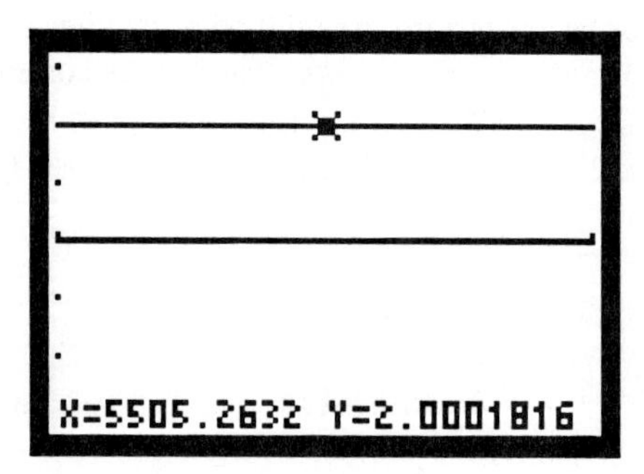

FIGURE 3.51

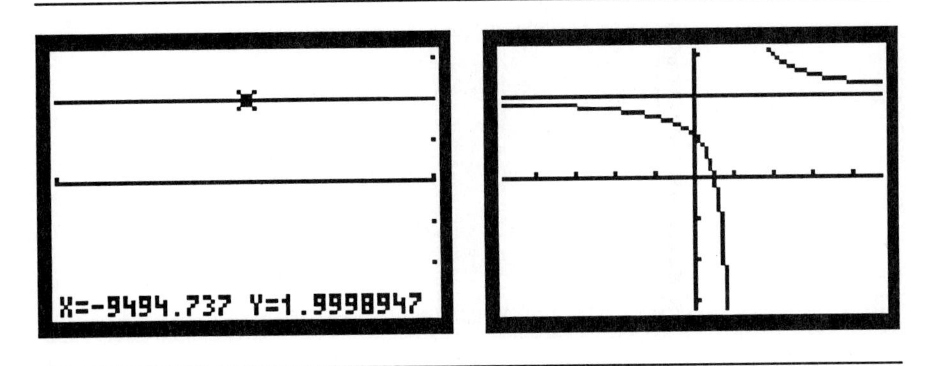

FIGURE 3.52

FIGURE 3.53

and therefore there is no value for x that makes the expression $\frac{1}{x} + 2 = 2$ true. This means that the graph of the function $y = \frac{1}{x} + 2$ will never cross the line $y = 2$. By making x a very large positive or negative number, we can make the function as close to a value of 2 as we wish; however, the value will never equal exactly 2.

This horizontal behavior is the same as we discovered at a vertical asymptote. For the function $y = \frac{1}{x} + 2$, the line $y = 2$ is a **horizontal asymptote.** Similarly, for the function $y = \frac{1}{x}$, the line $y = 0$ is a horizontal asymptote. In general, for the rational function $y = \frac{a}{x} + b$, where b is the vertical shift factor, the line $y = b$ is a horizontal asymptote.

Figure 3.53 shows the graphs of $y = \frac{1}{x-1} + 2$ and $y = 2$ in the window [-4.8, 4.7] by [-3.2, 3.1]. This rational function has an infinite discontinuity at the value $x = 1$, where the denominator becomes zero. As with other families of functions, the value added or subtracted from the x variable was related to the horizontal shift factor.

Use the **Line** command to draw the vertical asymptote at $x = 1$ (see Fig. 3.54). The point where the horizontal and vertical asymptotes cross, (1, 2), is the "center" of the shifted function. ◊

APPLICATION EXPLORATION: Tim Blake has to prepare his car for a cold Michigan winter by changing the antifreeze. The directions on the Summit Antifreeze/Coolant container specify that the mixture of antifreeze and water must be between 50% and 70% concentration of antifreeze. Tim has 15 liters of 30% antifreeze. He wants to add pure antifreeze to this 30% mixture to bring the concentration up to between 50% and 70%. What are the minimum and maximum amounts of antifreeze Tim

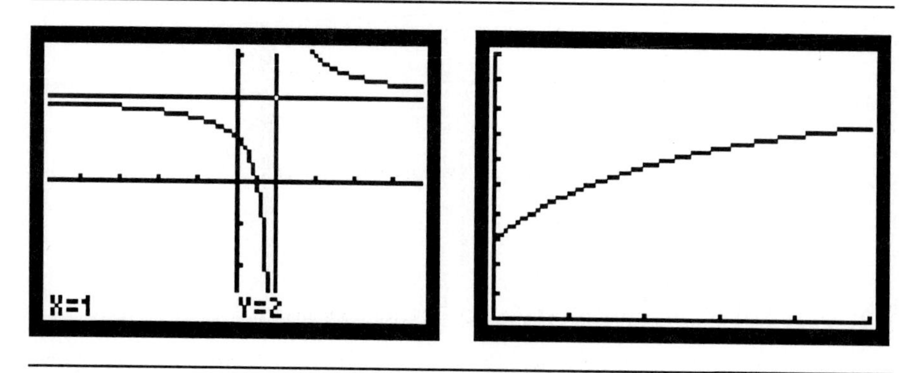

FIGURE 3.54

FIGURE 3.55

should add to the mixture to be within the manufacturer's guidelines? The radiator of Tim's car has a capacity of 30 liters. What is the concentration of this amount of mixture?

Concentration is given by the ratio $\frac{\text{amount of antifreeze}}{\text{total amount of liquid}}$. The amount of pure antifreeze in 15 liters of a 30% mixture is given by the product .30(15). Let x represent the amount of pure antifreeze added to the mixture. Concentration, as a function of x, is given by the rational function

$$C(x) = \frac{.30\,(15) + x}{15 + x}.$$

Concentration, represented on the y-axis, is a value between 0 and 1. The variable x represents the amount of antifreeze added. In the problem situation, the amount of antifreeze added is greater than zero. Figure 3.55 shows the graph of this rational function in the window [0, 25] by [0, 1]; this is the graph of the problem situation. Figure 3.56 shows the graph of this rational function in the window [-25, 25] by [-10, 10]; this is the complete graph of the function.

The mathematical question we need to answer for this problem situation is

$$0.5 \le \frac{.30\,(15) + x}{15 + x} \le 0.7.$$

Represent this inequality graphically by graphing the boundary conditions as $\mathbf{Y}_2 = .5$ and $\mathbf{Y}_3 = .7$. Figure 3.57 shows the graphical representation of this inequality in the viewing window [0, 25] by [0, 1]. The TRACE cursor is positioned near the lower bound of the manufacturer's specifications. Using zoom in, $x = 6$ when the concen-

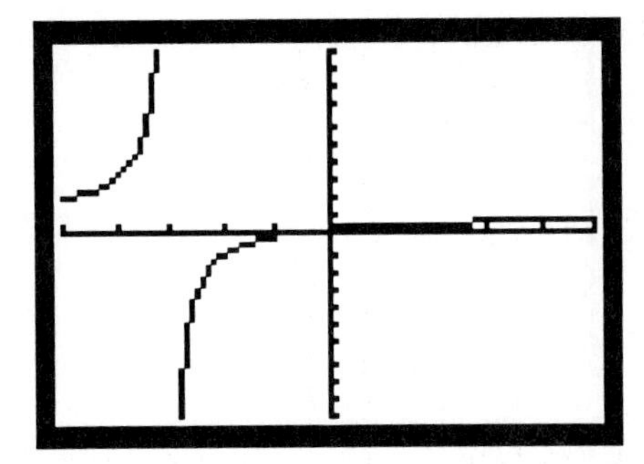

FIGURE 3.56

tration is 50%. This means that Tim should add 6 liters of antifreeze to the mixture to bring the concentration up to 50%.

Figure 3.58 shows the upper bound of the recommended mixture at the point (20, .7); adding 20 liters of antifreeze brings the concentration of the mixture up to 70%.

Since Tim started with 15 liters of mixture, adding 20 liters of pure antifreeze will make 35 liters of mixture. This is more than he needs. Since Tim only needs 30 liters for his car, he only needs to add 15 liters of antifreeze to the original 15 liters of 30% mixture to get exactly 30 liters of the new mixture. But, if he adds only 15 liters of pure antifreeze, will the concentration of the mixture be within the manufacturer's recommendations? Move the TRACE cursor along the graph of the rational function until the amount of pure antifreeze added is 15 liters. Figure 3.59 shows the TRACE cursor at the point (15, .65), indicating a concentration of 65% if 15 liters of pure antifreeze are added.

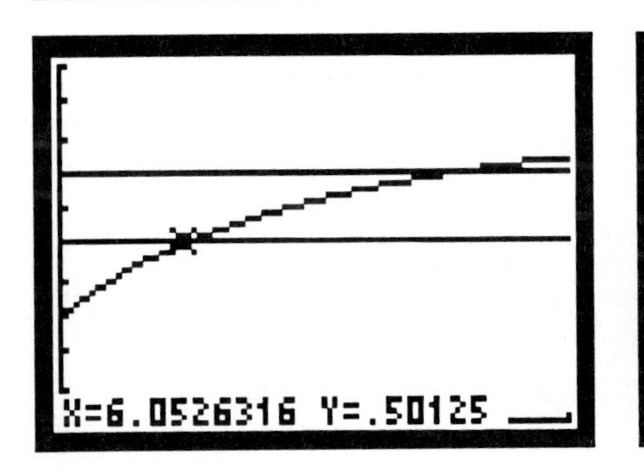

FIGURE 3.57

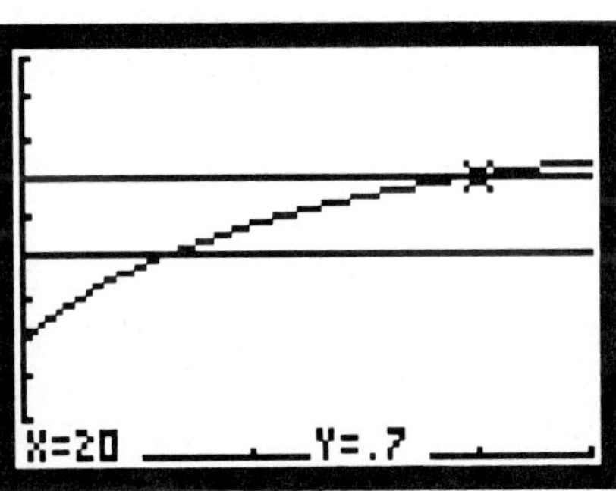

FIGURE 3.58

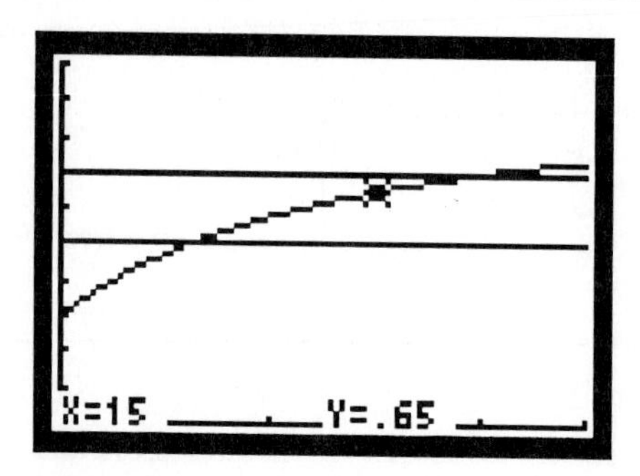

FIGURE 3.59

For Advanced Algebra Students

We can manipulate the original rational function $C(x) = \dfrac{.30(15) + x}{15 + x}$ into a form that will help us analyze the characteristics of the function and its graph. Rearrange the function:

$$C(x) = \frac{.30(15) + x}{15 + x} = \frac{x + 4.5}{x + 15}.$$

Perform the indicated long division:

$$\begin{array}{r|l} & 1 + \dfrac{-11.5}{x+15} \\ \hline x + 15 & x + 4.5 \\ & \underline{x + 15} \\ & -11.5 \end{array}$$

Then

$$C(x) = \frac{.30(15) + x}{15 + x} = 1 + \frac{-11.5}{x + 15}.$$

By rearranging,

$$C(x) = \frac{.30(15) + x}{15 + x} = -11.5\left(\frac{1}{x + 15}\right) + 1.$$

This form indicates the graph of this function is the graph of $y = \frac{1}{x}$ stretched by a factor of 11.5, rotated around the x-axis, shifted horizontally -15 units, and shifted vertically 1 unit. There is a vertical asymptote at $x = -15$ and a horizontal asymptote at $y = 1$. Refer back to Fig. 3.56 for a complete graph of this function. ◊

Problems

1. Algebraically explain the behavior of the graph of the function $y = \frac{1}{x}$:

a. Why is the function undefined at zero?

b. What happens when x is close to zero, but positive? Negative?

c. What happens when x gets large but is positive? Negative?

d. Use TRACE to see if the function ever reaches zero.

2. Graph the following on paper by transforming $y = \frac{1}{x}$, and check by graphing on the calculator.

a. $y = \frac{1}{x} - 2$ **b.** $y = \frac{1}{x-2}$

c. $y = \frac{2}{x}$ **d.** $y = \frac{-3}{x+3} + 2$

3. Not all rational functions have points of discontinuity or vertical asymptotes. Algebraically explain why the vertical asymptote is at $x = 3$ for the rational function $y = \frac{1}{x-3}$. From this, give the vertical asymptote(s) for the following and check graphically. If none, explain why.

a. $y = \frac{3}{2x-3}$ **b.** $y = \frac{x}{x+1}$

c. $y = \frac{1}{x^2-1}$ **d.** $y = \frac{1}{x^2+1}$

4. Graphically explore what happens when we take the reciprocal of some of our favorite functions we have studied so far. Can you explain the behavior algebraically?

a. $y = \frac{1}{x^2}$ **b.** $y = \frac{1}{x^3}$

c. $y = \frac{1}{x^2+x-2}$ **d.** $y = \frac{1}{x^2+2x+1}$

e. $y = \frac{1}{|x|}$

5. Use a graph to convince yourself that the following "cancellation" of crossing out the 3's is not correct:

$$\frac{\not{3}}{x+\not{3}}=\frac{1}{x}$$

6. For the following pairs of equations, examine the algebraic solution, and check the results by graphing the two original equations separately. Explain any differences in the algebraic and graphical solutions.

a. $y=x-\frac{1}{x}$ and $x^2-1=y$

$$x-\frac{1}{x}=x^2-1$$
$$\frac{x^2-1}{x}=x^2-1$$
$$x^2-1=x(x^2-1)$$
$$1=x$$

b. $y=x$ and $y=\frac{1}{1+\frac{1}{x}}$

$$x=\frac{1}{1+\frac{1}{x}}$$
$$x=\frac{x}{x+1}$$
$$x^2+x=x$$
$$x^2=0$$
$$x=0$$

7. Examine the following equations both algebraically and graphically to decide whether they are true for all x, some x, or no values of x. If possible, alter the equation so it is true for all x.

a. $\frac{1}{x}+\frac{1}{2}+\frac{2}{x+2}$

b. $\frac{2}{x}+\frac{3}{x-1}=\frac{6}{x(x-1)}$

c. $\frac{1}{2}-\frac{1}{x}=\frac{x-2}{2x}$

8. The acceleration of an object is given by the formula: $a = 2d/t^2$, where d is the distance traveled in meters, and t is the time in seconds. Amanda is on the track team and runs the 100 meter dash.

a. Write and graph a function that relates her acceleration and time in a 100 meter race.

b. What values make sense in this problem? Why is the function decreasing?

c. What acceleration does she need to run the race in 10 seconds? 12 seconds?

d. How fast does she run 100 meters if her acceleration is .888 m/sec^2?

9. Jason is twirling a 1-kg object at the end of a rope at a speed of 10 m/sec. To answer the following questions, use the fact that the force of an object being twirled around in a circular motion is given by the formula $F = mv^2/r$, where m is the mass, v is the velocity, and r is the distance (radius) the object is away from you.

a. What equation gives the force being exerted as Jason lets out the rope?

b. If Jason lets out 5 ft of rope, then what is the force?

c. If the force being exerted is 100 kg/sec, then how much rope has Jason let out? If the force is 200 kg/sec, how much rope is out?

d. Would you rather be struck by an object tied to a long or short rope if it were twirled at the same speed?

10. Dr. Frankenstein's assistant Igor is playing with 40 ounces of a 30% acid solution, but needs something stronger for the Doctor's experiments. How much pure acid should Igor add to make a solution that is 70% acid? Solve the equation both algebraically and graphically.

11. Three times a week Juan works out by riding his bike over a 50-mile course. The first leg of his course is 21 miles long, the second leg is 14 miles, and the third leg is 15 miles. On one particular day, Juan noticed that the wind affected his normal speed. On the first leg, he rode 8 miles per hour slower than normal. On the second leg he was able to pedal 5 miles per hour faster. On the third leg he rode his normal speed. What is Juan's normal speed if he completed the 50 mile course in 3 hours? What would Juan's normal speed be if he completed the course in 4 hours? In 2 hours? Solve these equations both graphically and algebraically.

12. Graph $y = x + \frac{1}{x}$ in a friendly window, and explain why the sum of a number and its reciprocal is always greater than 2 or less than -2. Prove this algebraically.

13. Graph the following rational functions in a friendly integer window, and use the TRACE cursor to find integer values for x that give an integer value for y. Confirm your findings numerically and explain.

    a. $y = \dfrac{x}{x-1}$ **b.** $y = \dfrac{x^2}{x-1}$

14. We graphed $y = x + \frac{1}{x}$, $y = x^2 + \frac{1}{x^2}$ in an earlier section and noticed a pattern concerning the y values of the turning points for the sum of x^n and its reciprocal. Graph the functions $y = ax^3 + \frac{b}{x^3}$, $y = ax^4 + \frac{b}{x^4}$ etc., to see if this pattern involving a and b is true for higher degree functions. What effect does an odd or even degree have on the symmetry of the graph?

15. Graph the general function $y = \frac{a}{x^2 + b}$ for different values of a and b to discover how they affect the curve. How can we numerically find the y-intercept? Why are there no vertical asymptotes?

16. Explore the function $y = \frac{x^n}{x^2 + 1}$ for different values of n and explain algebraically any patterns you see related to other polynomials we have studied. How does an odd or even value for n affect the graph? Why do they call these RATIOnal functions?

17. Graph the following rational functions. Explore what happens as x gets large positive or negative. Zoom out using **XFact = 2** and **YFact = 2.** Explain what happened to the "local" behavior. Do you recognize any shapes like a polynomial you have seen before? Can you explain the behavior algebraically? Is there a reason why they are called RATIOnal functions?

    a. $y = \dfrac{x-4}{x^2 + x - 2}$ **b.** $y = \dfrac{x^3 + 1}{x^2 + 1}$ **c.** $y = \dfrac{x}{x-1}$

18. The vertex (turning point) of a general quadratic function is the point $\left(\frac{-b}{2a}, c - \frac{b^2}{4a}\right)$. Is there a general expression for the turning point of a quadratic function divided by x? Graph the function $y = \frac{(x+a)(x+b)}{x}$ for different values of a and b, and find the turning point by zooming in and using TRACE. Compare your answers with the general point $(\sqrt{ab}, (\sqrt{a} + \sqrt{b})^2)$.

19. The vertex (turning point) of a general quadratic function is the point $\left(\frac{-b}{2a}, c - \frac{b^2}{4a}\right)$. Explore the relationship between the turning points of the graphs of the standard quadratic function $y = ax^2 + bx + c$ and the function $y = \frac{a}{x^2} + \frac{b}{x} + c$ for the same values of a, b, and c. Confirm the coordinates of the turning points graphically and numerically.

FOR ADVANCED ALGEBRA STUDENTS

20. Create and graph your own rational functions to investigate the following conditions. Since the degree of each polynomial in the numerator and the denominator is a positive whole number, the degrees are either greater than, less than, or equal to each other.

a. How can we find the y-intercept algebraically?

b. How can we find the x-intercept algebraically?

c. How can we find any possible vertical asymptotes algebraically? When there is a vertical asymptote, what happens to the y values as x gets closer to the undefined value? Under what conditions would a rational function not have any vertical asymptotes? Give an example.

d. When the degree of the numerator is larger than the degree of the denominator, what happens to the y values as x gets larger? Explain this numerically. (*Hint*: Zoom out.) Is there a polynomial that acts as an end-behavior model for the rational function? Can these be found more exactly algebraically?

e. When the degree in the denominator is larger than the degree of the numerator, what happens to the y values as x gets larger? Explain this numerically. This end-behavior model is called a horizontal asymptote. Can the function cross the horizontal asymptote? Explain with an example.

f. When the degree of the numerator and denominator are equal, what happens to the y values as x gets larger? Explain this numerically. This end-behavior model is also a horizontal asymptote.

21. You may have the impression that whenever a function is undefined, it has a vertical asymptote. Graph $y = \frac{x^3 - 1}{x - 1}$. Is there a vertical asymptote for this function? Make a table of values to investigate the behavior of this function near $x = 1$. Graph the function in the friendly window [-9.6, 9.4] by [-6.4, 6.2] so that we can TRACE to $x = 1$. What does y equal? Can you see the "hole" in the graph? Graph the numerator and denominator separately to investigate the zeros of these polynomial functions. Explain the pattern.

Perform the same analysis on the following functions. In addition, factor the numerators to find any common factors. Make and test a conjecture about the type of discontinuities and common factors in the numerator and denominator. How does this connect to the graphical interpretation of discontinuities?

What functions do these functions behave like for all values of x (except where they are undefined)? Besides the following functions create other rational functions that have "holes" instead of "poles" (asymptotes).

a. $y = \dfrac{x^2 - 3x + 2}{x - 2}$ **b.** $y = \dfrac{2x^2 + x - 6}{x + 2}$

c. $y = \dfrac{x^3 + 1}{x + 1}$ **d.** $y = \dfrac{x + 1}{x^2 - 1}$

22. Solve the following both algebraically and graphically. Explain any differences.

a. $3x = \dfrac{9}{x} + 26$ **b.** $\dfrac{4}{x-1} = \dfrac{5}{2x-2} + \dfrac{3x}{4}$

23. a. Solve the following system of equations algebraically:

$xy + x - y - 3 = 0$ and $xy - x - y - 1 = 0$.

b. Solve both of the original equations for y in terms of x and graph them to examine the problem. Use the TRACE cursor to find the distance between the curves for each value of x. What does it mean if they are the same distance apart at every x value?

c. Perform the indicated division on the function in part (b) and rearrange the results to show algebraically what you discovered graphically in part (b).

d. Consider the following system of equations:

$$y = \frac{x+6}{x-3} - \frac{12}{x+1} \qquad y = \frac{36}{x^2 - 2x - 3}.$$

How many roots does it have? Try solving it algebraically before doing so graphically.

24. Solve the inequality $\dfrac{2x+1}{1-2x} < 1$ both algebraically and graphically.

25. The function $y = \dfrac{8a^3}{x^2 + 4a^2}$ is called the Witch of Agnesi, and it is formed by having a fixed circle with its center at the point $(0, a)$ and passing through the origin. A horizontal tangent $y = 2a$ is drawn at the top of the circle. A rotating ray is drawn outward from the origin through the circle at point C and the tangent line at point T (see Fig. 3.60). The points of the curve, $P(x, y)$, are made up of the x-coordinate of the point T and the y-coordinate of the point C. As the ray is rotated the curve is formed by the locus of points of the two coordinates. What effect does changing the value of a have on the shape of this curve? Can you graph the circle and the tangent line on the same screen with the Witch of Agnesi curve?

26. One hundred pounds of cucumbers that are 95% water are dried in a food dehydrator until their water content is 90%.

a. What is the weight of the dried cucumbers?

b. Make a table that shows the weight of the dried cucumbers for various percentages of water content between 95% and 50%.

c. Write a function that shows the relationship between the weight of the cucumbers and the percentage of water present in them. Let x represent the percent of water removed

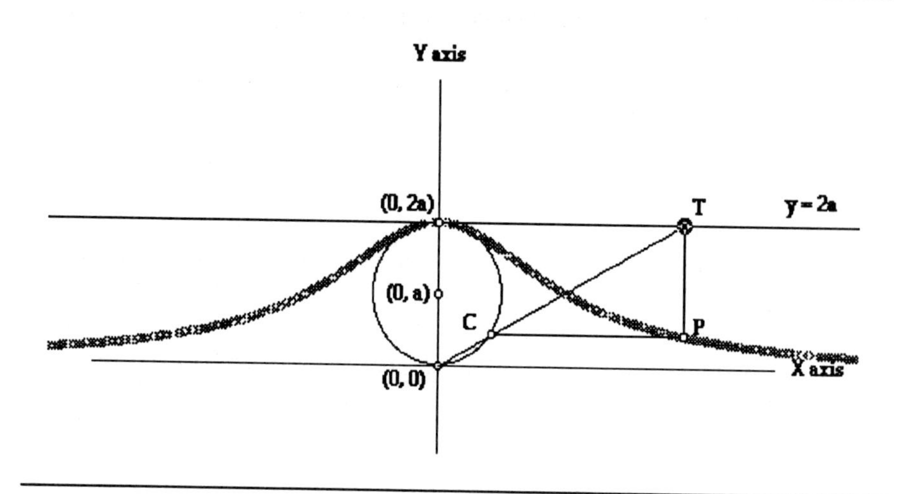

FIGURE 3.60
Locus of points forming the Witch of Agnesi curve

by the dehydrator and let y represent the resulting weight of the cucumbers. (*Hint*: Remember, 5 pounds of the original batch of cucumbers are the non-water part of the weight; this remains unchanged throughout the drying process.)

d. Draw a graph showing this relationship, and explain the shape of the graph in terms of the problem situation.

3.3 Exploring Radical Functions

In this section we will explore functions that involve **radicals**, for example,

$$y=\sqrt{x}, \qquad y=2+5\sqrt[3]{x-4}, \text{ or } \qquad y=\pm\sqrt{9-x^2}.$$

The radical symbol $\sqrt{\ }$ stands for the **square root,** and the radical symbol $\sqrt[3]{\ }$ stands for the **cube root** of the expression written under it. In general, the radical symbol $\sqrt[n]{\ }$ stands for the nth root of the expression, called the **radicand**; n is called the **index** of the radical. When $n = 2$, indicating a square root, the index is not written.

Radical expression can also be represented by fractional exponents. For example,

$$\sqrt{x+3}=(x+3)^{1/2}, \sqrt[4]{2x-x^2-x^3}=(2x-x^2-x^3)^{1/4}, \sqrt{x^3}=x^{3/2}.$$

These alternate forms using fractional exponents will be very useful ways to represent radical expressions.

EXPLORATION 1: Find the following values numerically with the TI-81:

$\sqrt{2}$ = ______________________ , $\sqrt{4}$ = ______________________ ,

- $\sqrt{10}$ = ______________________ , $\sqrt{-9}$ = ______________________ .

Figure 3.61 shows the first three calculations. Figure 3.62 shows the result of the final calculation. A **MATH** error has occurred.

When we find the square root of 4, we find the solution to the equation $x^2 = 4$. According to the calculator, the solution is 2 since $2^2 = 4$. What is the solution to the equation $x^2 = -9$? Stated in words, what number squared is equal to -9? The answer is no real value. The **MATH** error occurred because there is no real number that is the square root of a negative number. Later in your study of mathematics you will learn that the square root of a negative number is an **imaginary number.**

When the calculator returns the value 2 for the $\sqrt{4}$, we say that 2 is the solution to the equation $x^2 = 4$. However, it is also true that $(-2)^2 = 4$, so why not say $\sqrt{4} = -2$?

To solve this dilemma, we resort to a definition. By definition, the **principal square root** of a positive real number is positive. Your TI-81 always returns the principal square root of a value.

Calculate the value of $\sqrt[3]{8}$ and $\sqrt[3]{-8}$ (use option **4:** $\sqrt[3]{}$ from the MATH menu; see Fig. 3.63). This means that the cube roots of both positive and negative numbers are real numbers. In this example, $2^3 = 8$ and $(-2)^3 = -8$.

To find other roots of numbers, we must use fractional exponents. Figure 3.64 shows the calculation of $\sqrt[7]{-5}$, $\sqrt[4]{12}$, and $\sqrt[10]{9.5}$ using fractional (or decimal) exponents.

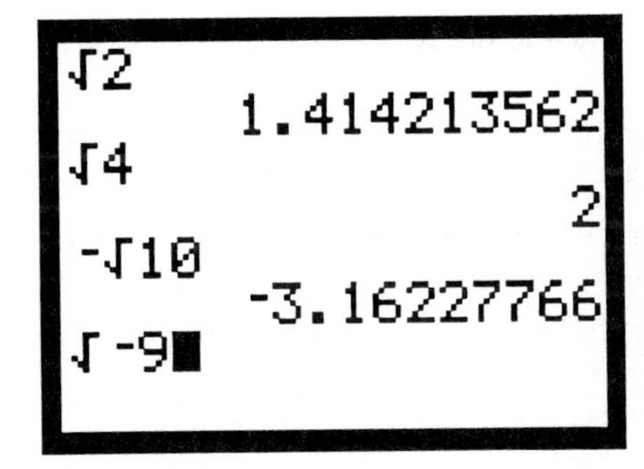

FIGURE 3.61

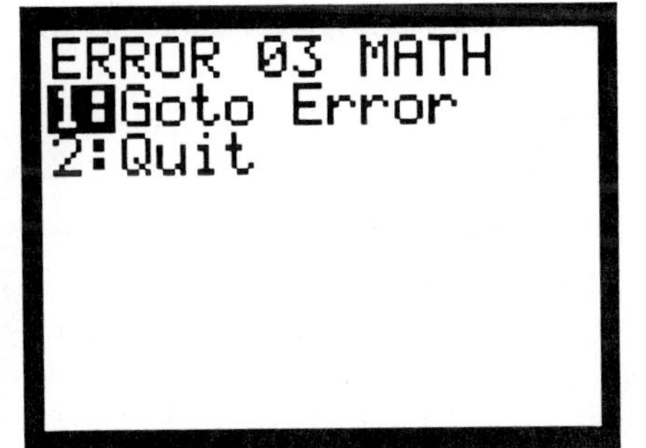

FIGURE 3.62

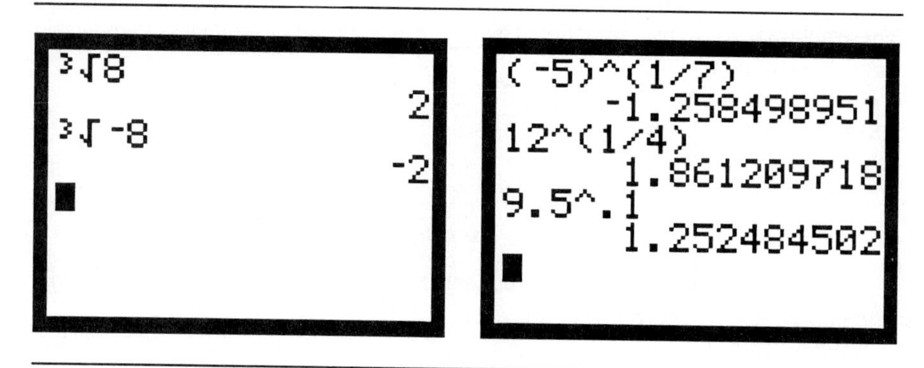

FIGURE 3.63

FIGURE 3.64

When is the root of a negative number a real value? When do you get a **MATH** error? In general, when the index of the radical is even, the root of a negative number is undefined; and when the index is odd, the root of a negative number is defined. To see this better, graph the function $y = \sqrt{x}$ and $y = \sqrt[3]{x}$ and compare the x values where the functions are defined. Figure 3.65 shows the graph of $y = \sqrt{x}$ and Fig. 3.66 the graph of $y = \sqrt[3]{x}$ in the [-4.8, 4.7] by [3.2, 3.1] viewing window.

The graph of $y = \sqrt{x}$ is defined only for positive values of x. The graph of $y = \sqrt[3]{x}$ is defined for all values of x. Figure 3.67 shows the graphs of $y = \sqrt{x}$, $y = \sqrt[4]{x}$, and $y = \sqrt[8]{x}$. Figure 3.68 shows the graphs of $y = \sqrt[3]{x}$, $y = \sqrt[5]{x}$, and $y = \sqrt[11]{x}$.

Use the TRACE cursor to find any common points for the sets of graphs in Figs. 3.67 and 3.68. Explain why these points appear on all the graphs. ◊

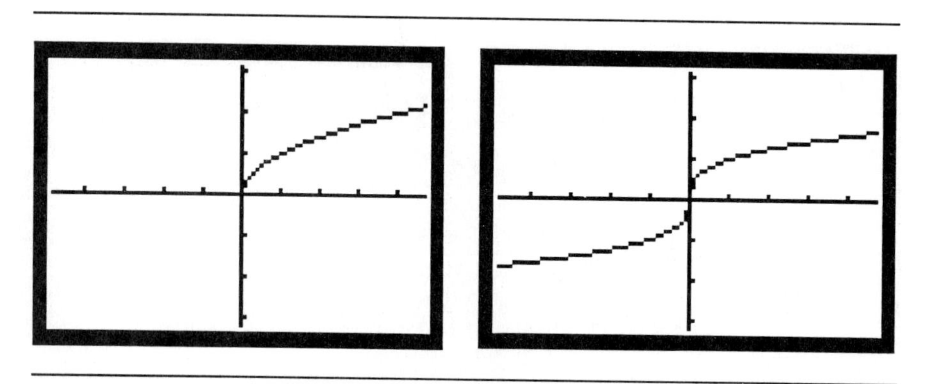

FIGURE 3.65

FIGURE 3.66

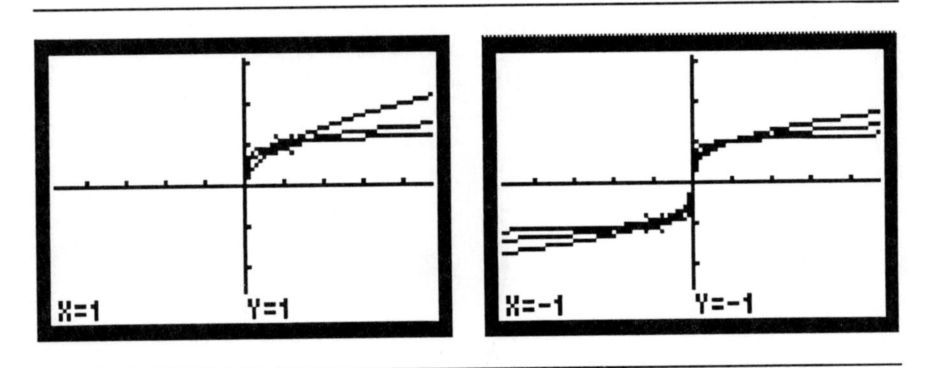

FIGURE 3.67

FIGURE 3.68

EXPLORATION 2: Compare the graphs of $y = \sqrt{x}$, $y = -\sqrt{x}$, and $y = \sqrt{-x}$. Explain the position of these graphs in the plane.

Figure 3.69 shows the graphs of $y = \sqrt{x}$ and $y = -\sqrt{x}$ in the [-4.8, 4.7] by [-3.2, 3.1] viewing rectangle. The negative sign has the effect of rotating the parent function around the x-axis, just as we have seen with other families of functions.

Figure 3.70 shows the graphs of $y = \sqrt{x}$ and $y = \sqrt{-x}$ in the same basic friendly window. The negative sign inside the radical of $y = \sqrt{-x}$ has the effect of rotating the function around the y-axis. Use the TRACE cursor to explain the position of the graph.

When x represents a negative number, then $-x$ is a positive number; since the square root of a positive number is a real number, the graph plots for the negative values of x. When x represents a positive number, $-x$ is negative and the square root is not defined. ◊

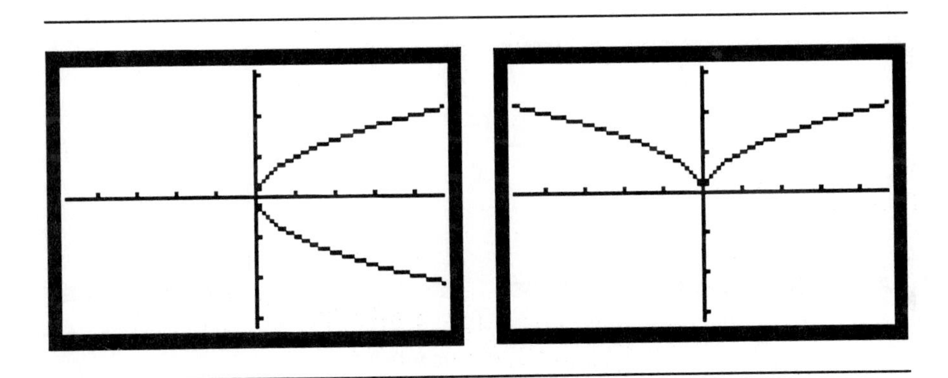

FIGURE 3.69

FIGURE 3.70

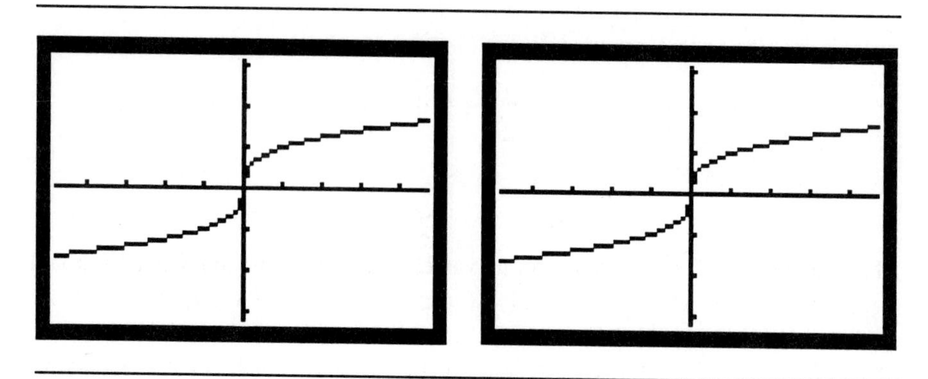

FIGURE 3.71
Graph of $Y_1 = \sqrt[3]{x}$

FIGURE 3.72
Graph of $Y_2 = X$^(1/3)

EXPLORATION 3: Explore the graph $y = \sqrt[3]{x}$ compared with the graph of $y = x^{1/3}$, and the graph of $y = \sqrt[3]{x^2}$ compared with $y = x^{2/3}$. Explain why the graphs appear as they do, and use alternate methods to define the functions to produce the correct graphs.

Figure 3.71 shows the graph of $y = \sqrt[3]{x}$ and Fig. 3.72 the graph of $y = x^{1/3}$ in the [-4.8, 4.7] by [-3.2, 3.1] viewing window. These functions are defined as

$$\mathbf{Y_1 = \sqrt[3]{X}} \quad \text{and} \quad \mathbf{Y_2 = X \wedge (1/3).}$$

Figure 3.73 shows the graph of the radical function $y = \sqrt[3]{x^2}$, and Fig. 3.74 shows the graph of the exponential function $y = x^{2/3}$. These two functions are defined as

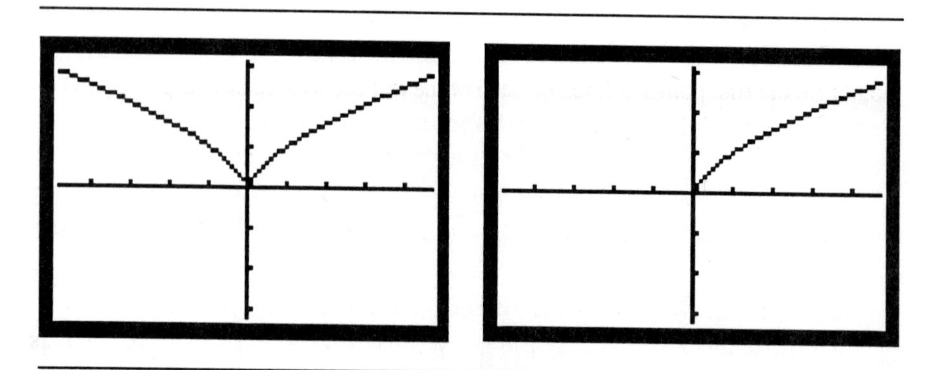

FIGURE 3.73
Graph of $Y_1 = \sqrt[3]{X^2}$

FIGURE 3.74
Graph of $Y_2 = X$^(2/3)

$\mathbf{Y_1 = \sqrt[3]{X^2}}$ and $\mathbf{Y_2 = X\text{^}(2/3)}$.

If the equation $\sqrt[3]{x^2} = x^{2/3}$ is an identity, then why are the two graphs different? Which one is correct?

Previously we discovered that the cube roots of both positive and negative numbers are defined, as shown by the graph of $y = \sqrt[3]{x}$ in Fig. 3.71. The radicand, x^2, of the function $y = \sqrt[3]{x^2}$ is positive for all values of x, so the cube root function should be defined for all values of x. Rational exponents like $x^{2/3}$ are difficult for any numerical calculating device to interpret correctly. As we saw in Figure 3.72, the TI-81 interprets the function $x^{1/3}$ correctly. When using a rational exponent with a numerator other than 1, care must be taken to be certain the correct graph is produced. Often we must use alternate definitions for rational exponents. For example, the following equations all represent the function $y = \sqrt[3]{x^2}$:

$\mathbf{Y_1 = \sqrt[3]{X^2}}$,
$\mathbf{Y_2 = X\text{^}(2/3)}$,
$\mathbf{Y_3 = (X\text{^}(1/3))^2}$,
$\mathbf{Y_4 = (X^2)\text{^}(1/3)}$.

$\mathbf{Y_1}$, $\mathbf{Y_3}$, and $\mathbf{Y_4}$ produce correct graphs; $\mathbf{Y_2}$ does not. These alternate definitions can be created by manipulating the fractional exponent:

$$x^{2/3} = (x^2)^{1/3} = (x^{1/3})^2.$$ ◇

EXPLORATION 4: Based on transformations done previously, predict the motions that will transform the graph of $y = \sqrt{x}$ into the graphs of $y = \sqrt{x-2}$ and $y = \sqrt{3x-6}$.

Figure 3.75 shows the graphs of $y = \sqrt{x}$ and $y = \sqrt{x-2}$ in the [-9.6, 9.4] by [-6.4, 6.2] viewing window. Use the TRACE cursor to show the point (0, 0) on the graph of $y = \sqrt{x}$ has been translated to the point (2, 0) on the graph of $y = \sqrt{x-2}$. These two points are the first points defined on each of the respective graphs. The point (1, 1) on $y = \sqrt{x}$ has been translated to the point (3, 1) on the graph of $y = \sqrt{x-2}$ by the horizontal translation of 2 units. Other points can be mapped in the same way.

Figure 3.76 shows the graphs of $y = \sqrt{x}$, $y = \sqrt{x-2}$, and $y = \sqrt{3x-6}$ in the [-9.6, 9.4] by [-6.4, 6.2] viewing window. Why do the graphs of both $y = \sqrt{x-2}$ and $y = \sqrt{3x-6}$ seem to be translated 2 units to the right? Should the graph of $y = \sqrt{3x-6}$ be translated 6 units to the right? Some algebraic manipulation is necessary to explain the position of the graph. Factor the radicand, and write the result as the product of two radicals:

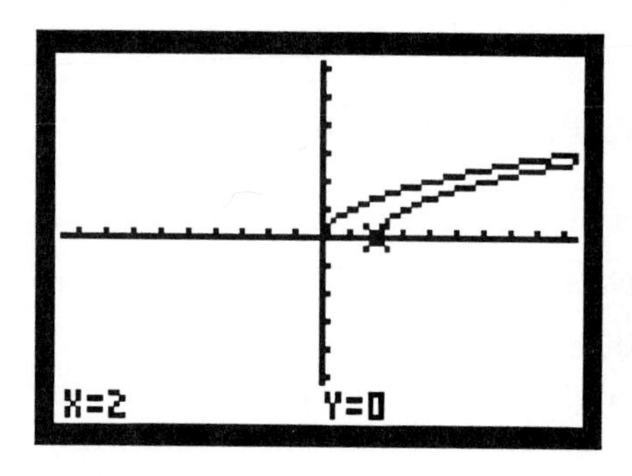

FIGURE 3.75

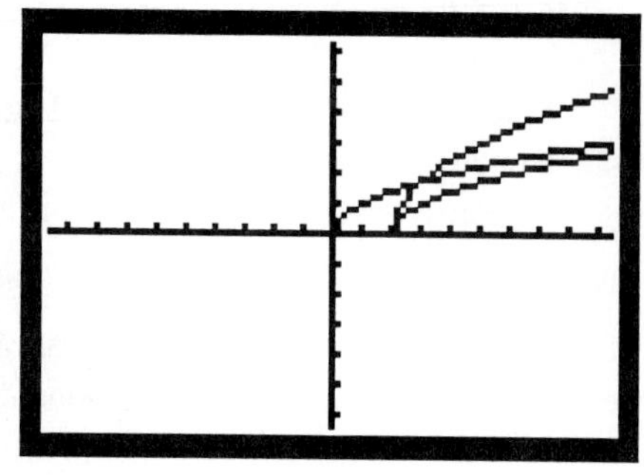

FIGURE 3.76

$$y = \sqrt{3x - 6}$$
$$y = \sqrt{3\,(x - 2)}$$
$$y = \sqrt{3}\,\sqrt{x - 2}$$

With the coefficient of x now equal to 1, we can predict a horizontal shift of 2 units to the right, as the graph shows. The factor $\sqrt{3}$ is a vertical stretch factor as seen in Fig. 3.76. In general, when predicting the horizontal shift factor, the coefficient of x must equal 1. ◊

APPLICATION EXPLORATION: Pat Milford is an engineer for ProCom Microwave Transmission Co. He is estimating the amount of support cable needed for a new 350-ft transmission tower to be built next month. These microwave towers are supported by three sets of three cables with each set anchored at the same pad. Figure 3.77 shows a sketch of a typical installation. The distance d to the anchor pads must be one-half the height of the tower and is the same for all three pads. The cables are attached to the top of the tower and at points $\frac{1}{3}$ and $\frac{2}{3}$ of the height.

Pat would like to write a function that would help him quickly estimate the various amounts of cable needed to support the tower. How can he do this?

The tower is perpendicular with the ground, forming right triangles with the cables as the hypotenuse of each triangle. The Pythagorean formula gives the length of the hypotenuse of a right triangle based on the length of the legs. The distance d to the anchor pads is established by the height of the tower, 350 ft. In general the length of a cable is given by the formula $l = \sqrt{h^2 + d^2}$, where h is the height to the anchor point on the tower and d is the distance to the anchor pad.

To graph this relationship, let x represent the height, h, and let y represent the length of the cable, l. The value of d will be a constant in the problem. The function will be

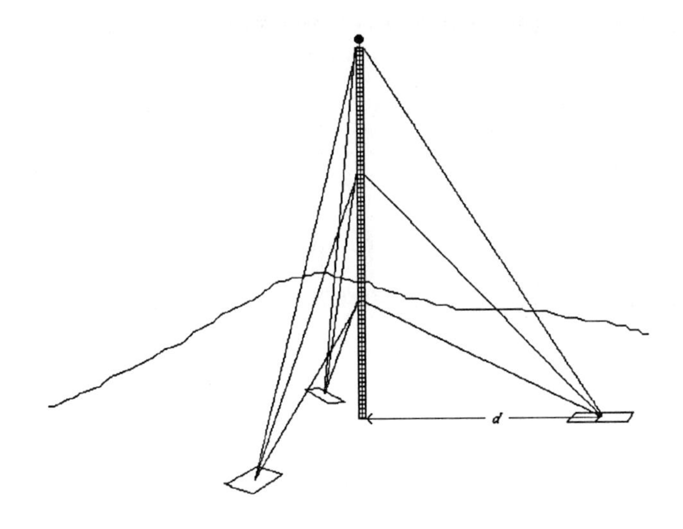

FIGURE 3.77
Microwave transmission tower installation

$$y = \sqrt{x^2 + \left(\frac{350}{2}\right)^2}\,.$$

The variable x represents different heights along the tower from 0 to 350 ft. The variable y represents the length of the cable; the longest cable will necessarily be longer than 350 ft. Why? Figure 3.78 shows the graph of this function in the viewing window [0, 380] by [0, 500].

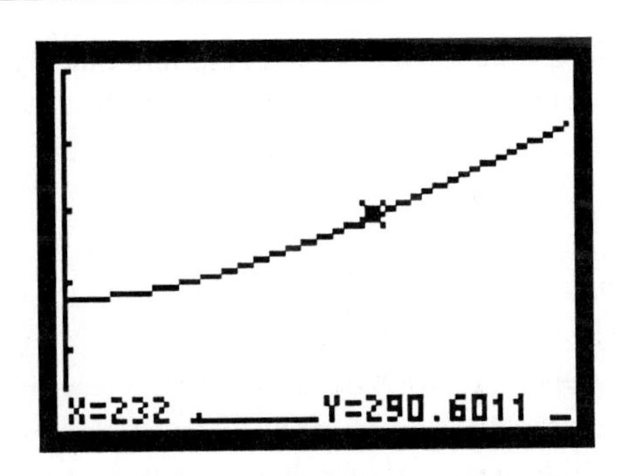

FIGURE 3.78

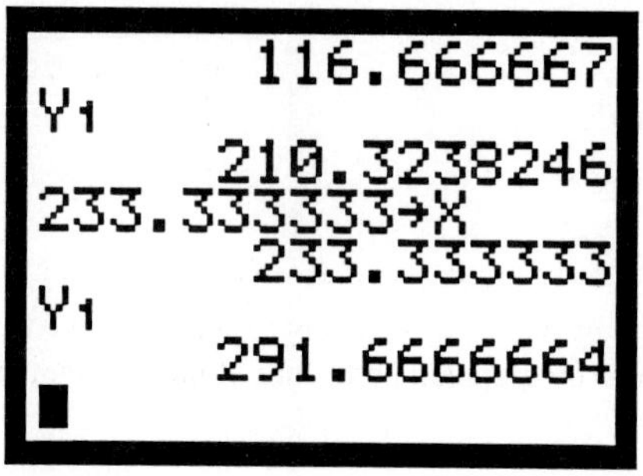

```
         116.666667
Y1
         210.3238246
233.333333→X
         233.333333
Y1
         291.6666664
■
```

FIGURE 3.79

The cables must be attached to the tower at 350/3 = 116.67 ft, at 2 × 350/3 = 233.33 ft, and at 350 ft. Use the TRACE cursor to estimate the amount of cable needed for one anchor pad: 211 + 292 + 391 = 894. Multiply this figure by 3 to get an estimate for the total job: 894 × 3 = 2682 ft of cable.

An alternate way to get accurate estimates for the length of the cable is to enter the various heights in the **X** memory register and then evaluate the function $\mathbf{Y_1}$. Figure 3.79 shows two of the cable lengths evaluated this way. ◊

Problems

1. Graph $y = \sqrt{x}$, and explain algebraically why it is only being graphed in the first quadrant. As x values get large, what happens to the y values? Why does the function increase so much slower than $y = x^2$, and even $y = x$?

2. Sketch the graph of the following functions by hand using transformations of $y = \sqrt{x}$, and then confirm using the graphing calculator.

   a. $y = \sqrt{x} + 1$ **b.** $= \sqrt{x-1}$

   c. $y = 5\sqrt{x}$ **d.** $y = \sqrt{-5x}$

   e. $y = -2\sqrt{x}$ **f.** $y = 2 + 3\sqrt{x-4}$

3. In a friendly window, graph and compare $y = x^2$ with $y = \sqrt{x}$ and $y = -\sqrt{x}$. What do you notice? Use the TRACE cursor to move from a value on $\mathbf{Y_1}$ to values on $\mathbf{Y_2}$ and $\mathbf{Y_3}$ to compare the coordinates. Add the graph $\mathbf{Y_4 = X}$ to help see the symmetry. Explain what you see algebraically.

4. Draw the graph of $y = \sqrt{1-x}$. What are the domain and range of this function? Explain what you see numerically. Explain the graph as a transformation of $y = \sqrt{x}$. Sketch the graph of $y = \sqrt{3-x}$ without the calculator. Explain in general how to sketch the graph of $y = \sqrt{a-x}$.